URBAN CLIMATE POLITICS: AGENCY AND EMPOWERMENT

Since the 1990s, a burgeoning literature has emerged on urban climate politics and governance. It is now evident that urban responses to climate change involve a diverse range of actors as well as forms of agency that cross traditional boundaries, and that have diverse consequences for (dis)empowering different social groups.

This book provides an overview of the forms of agency in urban climate politics, discussing the friction and power dynamics between them. Written by renowned scholars, it critically assesses the advantages and limitations of increasing agency in urban climate governance. In doing so, it sheds critical new light on the existing literature, advances the state of knowledge of urban climate governance, and discusses ways to accelerate urban climate action. With chapters building on case studies from across the world, it is ideal for scholars and practitioners working in the area of urban climate politics and governance.

JEROEN VAN DER HEIJDEN is Professor of Governance and Chair in Regulatory Practice at the School of Government, Victoria University of Wellington (New Zealand) and Honorary Professor at the School of Regulation and Global Governance, Australian National University (Australia). His research is concerned with innovations in urban climate governance in the Global North and Global South. Recent books include *Innovations in Urban Climate Governance* (Cambridge University Press, 2017) and *Governance for Urban Sustainability and Resilience* (2014).

HARRIET BULKELEY is Professor of Geography and holds a joint appointment at Durham University (UK) and Utrecht University (the Netherlands). Her research is concerned with the nature and politics of environmental governance, and she has particular expertise in the areas of climate change, energy, and urban sustainability. Recent books include *Transnational Climate Change Governance* (edited with L. B. Andonova et al.; Cambridge University Press, 2017) and *Cities and Climate Change* (2013). She currently convenes the H2020 NATURVATION project examining urban innovation with nature-based solutions for sustainable development.

CHIARA CERTOMÀ is Marie Skłodowska-Curie Research Fellow at Ghent University (Belgium) and an Affiliated Researcher at the Sant'Anna School of Advanced Studies (Italy). Her principal research topics include the politics of space and place, the effects of informal urban planning practices, and the role of new information and communication technologies in urban sustainability governance. Recent books include *Citizen Empowerment and Innovation in the Data-Rich City* (edited with M. Dyer, L. Pocatilu, and F. Rizzi; 2017) and *Postenvironmentalism: A Material Semiotic Perspective on Living Spaces* (2016).

The **Earth System Governance Project** was established in 2009 as a core project of the International Human Dimensions Programme on Global Environmental Change. Since then, the Project has evolved into the largest social science research network in the area of sustainability and governance. The Earth System Governance Project explores political solutions and novel, more effective governance mechanisms to cope with the current transitions in the socio-ecological systems of our planet. The normative context of this research is sustainable development; earth system governance is not only a question of institutional effectiveness, but also of political legitimacy and social justice.

The **Earth System Governance series** with Cambridge University Press publishes the main research findings and synthesis volumes from the Project's first ten years of operation.

Series Editor

Frank Biermann, Utrecht University, the Netherlands

Titles in print in this series

Biermann and Lövbrand (eds.), *Anthropocene Encounters: New Directions in Green Political Thinking*

van der Heijden, Bulkeley and Certomà (eds.), *Urban Climate Politics: Agency and Empowerment*

URBAN CLIMATE POLITICS

Agency and Empowerment

Edited by

JEROEN VAN DER HEIJDEN
Victoria University of Wellington and Australian National University

HARRIET BULKELEY
Durham University and Utrecht University

CHIARA CERTOMÀ
Ghent University and Sant'Anna School of Advanced Studies

CAMBRIDGE
UNIVERSITY PRESS

CAMBRIDGE
UNIVERSITY PRESS

University Printing House, Cambridge CB2 8BS, United Kingdom

One Liberty Plaza, 20th Floor, New York, NY 10006, USA

477 Williamstown Road, Port Melbourne, VIC 3207, Australia

314–321, 3rd Floor, Plot 3, Splendor Forum, Jasola District Centre, New Delhi – 110025, India

79 Anson Road, #06–04/06, Singapore 079906

Cambridge University Press is part of the University of Cambridge.

It furthers the University's mission by disseminating knowledge in the pursuit of education, learning, and research at the highest international levels of excellence.

www.cambridge.org
Information on this title: www.cambridge.org/9781108492973
DOI: 10.1017/9781108632157

First published 2019

Printed and bound in Great Britain by Clays Ltd, Elcograf S.p.A.

A catalogue record for this publication is available from the British Library.

Library of Congress Cataloging-in-Publication Data
Names: Heijden, Jeroen van der, 1977– editor. | Bulkeley, Harriet, 1972– editor. | Certomà, Chiara, 1979– editor.
Title: Urban climate politics : agency and empowerment / edited by Jeroen van der Heijden, Victoria University of Wellington; Harriet Bulkeley, University of Durham; Chiara Certomà, University of Ghent.
Description: Cambridge, United Kingdom ; New York, NY : Cambridge University Press, 2019. | Series: The earth system governance series | Includes bibliographical references.
Identifiers: LCCN 2018049979 | ISBN 9781108492973
Subjects: LCSH: Urban climatology. | Urban ecology (Biology) | Urban policy – Environmental aspects. | City planning – Environmental aspects.
Classification: LCC QC981.7.U7 U6325 2019 | DDC 307.76–dc23
LC record available at https://lccn.loc.gov/2018049979

ISBN 978-1-108-49297-3 Hardback
ISBN 978-1-108-73022-8 Paperback

Contents

Contributors

Fabiana Barbi is a research associate at the Centre for Environmental Studies and Research, University of Campinas Brazil.

Antje Bruns is a professor at the Governance and Sustainability Lab, Trier University, Germany.

Sara Burch is an associate professor at the Department of Geography and Environmental Management, University of Waterloo, Canada.

Peter P. J. Driessen is a professor at the Copernicus Institute of Sustainable Development, Utrecht University, the Netherlands.

Sarah Giest is an assistant professor at the Institute of Public Administration, Leiden University, the Netherlands.

David J. Gordon is an assistant professor at the Department of Politics, University of California Santa Cruz, USA.

Christopher Gore is an associate at the Department of Politics and Public Administration, Ryerson University, Canada.

Dries L. T. Hegger is an assistant professor at the Copernicus Institute of Sustainable Development, Utrecht University, the Netherlands.

Thomas Hickmann is a postdoctoral researcher at the Chair of International Politics, University of Potsdam, Germany.

Chris Höhne is a PhD candidate and research associate at the Institute of Political Science, Technical University of Darmstadt, Germany.

Sara Hughes is an assistant professor at the Department of Political Science, University of Toronto, Canada.

Markus Lederer is a professor at the Institute of Political Science, Technical University of Darmstadt, Germany.

François Mancebo is a professor at the Faculty of Humanities and Social Sciences, University of Reims Champagne-Ardenne, France.

Heleen L. P. Mees is an assistant professor at the Copernicus Institute of Sustainable Development, Utrecht University, the Netherlands.

Scott Morton Ninomiya is a PhD student at the Balsillie School of International Affairs, University of Waterloo, Canada.

James J. Patterson is an assistant professor at the Copernicus Institute of Sustainable Development, Utrecht University, the Netherlands.

Lauren Rickards is an associate professor at the School of Global, Urban and Social Studies, RMIT University, Australia.

Karsten Schulz is an assistant professor at the Data Research Centre, University of Groningen, the Netherlands.

Fee Stehle is a doctoral researcher at the Chair of International Politics, University of Potsdam, Germany.

Laura Tozer is a postdoctoral researcher in the Department of Geography, Durham University, UK.

Caroline J. Uittenbroek is a postdoctoral researcher at the Copernicus Institute of Sustainable Development, Utrecht University, the Netherlands.

Laura Valente de Macedo is a postdoctoral researcher at the São Paulo School of Management, Fundação Getulio Vargas, Brazil.

Nicolien van der Grijp is a senior researcher at the Institute for Environmental Studies, Free University of Amsterdam, the Netherlands.

Acknowledgements

The idea for this book began late 2016, on an Australian roadside, in an English home, and in an Italian office. On a trip from the Blue Mountains to Canberra, Jeroen had planned a video-call with Harriet to discuss some first ideas for an edited book within the Earth System Governance (ESG) Project's Harvesting Initiative. Not long after, in another video-call – albeit from a less scenic spot in Australia – Chiara joined in also.

The ESG Project was launched in 2009. It has become the largest social science research network in the area of governance and global environmental change. Over the years, it has expanded into a broad community of 300 global scholars, a dozen research centres, annual conferences, and a lively online presence. Now, close to a decade after its launch, it is time to harvest what has been achieved by ESG scholars – this is, in a nutshell, what the ESG Harvesting Initiative entails. This book does so for ESG work on urban climate transitions.

Many of the chapters in this book were first presented at the 2017 Lund Conference on Earth System Governance: "Allocation & Access in a Warming and Increasingly Unequal World." At this conference, Jeroen, Harriet, and Chiara held two workshops to discuss draft chapters and think critically about notions of increasing agency and contested empowerment in urban climate politics. We would like to thank the authors and audience for their participation in these workshops and for the valuable input they have given us and each other toward improving the various chapters in this book.

We would also like to thank the publishing team at Cambridge University Press, and in particular Emma Kiddle, for their support and help with the commissioning and production of this book. Many thanks also to Frank Biermann, Ruben Zondervan, and other ESG staff and faculty for coordinating the ESG Project's Harvesting Initiative. A word of gratitude also goes to the five anonymous reviewers for providing us with helpful suggestions to further improve the book.

Jeroen, Harriet, and Chiara would like to acknowledge support from the Australian Research Council (DECRA Award number DE15100511), the European Union's Horizon 2020 Research and Innovation programme (Marie Skłodowska-Curie Action, grant agreement 740191), the Netherlands Organisation for Scientific Research (VIDI Fellowship number 45111015), and the New Zealand Government Regulatory Practice Initiative (G-REG).

Bringing together thirteen chapters on contested issues such as agency and empowerment from more than twenty scholars in a single book volume sounded like a daunting task at first but has turned out to be a very pleasant process. Many thanks therefore also, finally, to the authors of this book for keeping to the deadlines we set and responding to our ongoing requests for sharpening the arguments they make in their respective chapters.

1

Promises and Concerns of the Urban Century

Increasing Agency and Contested Empowerment

JEROEN VAN DER HEIJDEN, HARRIET BULKELEY, AND CHIARA CERTOMÀ

Stressing the relevance of urbanization in social, economic, and environmental developments, the twenty-first century is frequently referred to as 'the first urban century' or simply 'the urban century' (Gilbert et al. 2009; Hall & Pfeiffer, 2013; Heynen, 2014; Nijkamp & Kourtit, 2013). The numbers speak for themselves: as of 2008 the world's urban population was larger than the world's rural population. By 2050, some 6 billion people are expected to live in cities and urban areas – twice as many as in 2000. Most rapid urbanization will take place in Asia and Africa, where urban populations are expected to grow from around 30 per cent at the start of the century to more than 50 per cent by mid-century (UN-HABITAT, 2016). Already more than 70 per cent of global gross domestic product (GDP) is generated in cities, and with increasing urbanization cities will become even more dominant in the world economy (McKinsey Global Institute, 2011; World Bank, 2009b). Some 70 per cent of global resources are consumed in cities (including energy and potable water) and they account for 70 per cent of global greenhouse gas emissions – mainly as a result of the high consumerist lifestyle that characterizes modern urban life (Dodman, 2009; UN, 2016).

The social, economic, and environmental gains and costs of urbanization are not spread equally across the world, however. Half of global GDP is generated in fewer than 400 cities predominantly in the global north (McKinsey Global Institute, 2011; World Bank, 2009b). Problems such as urban poverty, slums, air pollution, overabstraction of drinking water, heat waves, and flooding are more persistent in the Global South than in the Global North (Hughes, Chu, & Mason, 2018; Mitlin & Satterthwaite, 2013; Watson, 2009). And in both the Global North and the Global South, inequalities between the haves and the have-nots are vast and rapidly growing (World Bank Group, 2016). Yet, the ongoing urbanization and redevelopment of cities holds also much potential for reduced resource consumption and waste production, as well as opportunities for a more just and equal distribution of the yields and cost of economic and social development (Bulkeley, 2013; Castán

Broto, 2017; Hughes, 2017; Van der Heijden, 2017). In other words, not only can we *imagine* urban futures with greater well-being of individuals and societies, but also, given the scale of projected urbanization and redevelopment there are real possibilities to *achieve* urban futures with greater well-being of individuals and societies (Drydyk, 2013; Ibrahim & Alkire, 2007; Wright, 2010).

It is for these reasons that urban responses to climate change have become increasingly significant over the last decades and will remain critical to achieve equally distributed social and economic progress (Bulkeley & Betsill, 2003; Bulkeley, Castán Broto, & Gareth, 2012; Bulkeley, Edwards, & Fuller, 2014; Cohen, 2017; Van der Heijden, 2016). Reflecting this, a burgeoning literature has emerged on the politics and governance of urban futures, particularly in the area of climate change adaptation and mitigation. This literature has a strong focus on the opportunities and constraints of innovative and experimental policies and governance instruments that have emerged since the early 1990s. This work has positioned the 'urban' as an important arena for Earth System Governance and dominant urban actors – municipal governments and the transnational associations which they have formed – as critical agents of change in the transition towards a low carbon and resilient future. However, while other domains of environmental governance have come to explore the multiple actors involved in governing climate change and the range of forms of agency involved, within the field of urban studies of climate change there has been a more limited engagement with the diverse agents and novel forms of agency that have engaged in urban responses. Beyond city governments, state and regional public authorities have also been critical to the urban governing of climate change, and there is a growing involvement of non-governmental organizations, citizen collectives, transnational municipal networks, development banks, philanthropic organizations, and businesses.

As these new agents of change have started to engage with the urban governance of climate change, new questions concerning their roles and responsibilities are emerging. Central to these issues are matters of their legitimacy (to whom are such agents answerable?) and the extent to which they are able to empower citizens and communities to undertake transformative action for climate change. While it can be tempting to equate novel forms of collective action as more likely to support such transformative action, the extent to which such initiatives are truly accessible and which forms of community come to benefit need to be further reflected upon. Equally, neo-liberal drivers in areas such as smart cities might increase inequalities between citizens rather than reduce them, and through innovative urban climate governance interventions governments may bypass their constitutional limits and affect actors at scales or levels where they normally do not have jurisdiction. At the same time, while the move of corporations and financial institutions into urban climate governance might raise concerns about their motivations and transparency, there is the potential for such initiatives to leverage the forms of resources required to develop capacity and empower others to take action. We need to engage with these potentially contradictory dynamics if the full implications of urban climate governance are to be recognized.

In sum, it is now evident that urban responses to climate change involve a diverse range of actors as well as forms of agency that cross traditional boundaries, and which have diverse consequences for (dis)empowering different social groups – helping or hampering them to increase their well-being. Friction between novel forms of agency, new agents of change and (dis)empowerment is a missing focus in existing scholarship on urban climate futures. This edited book addresses this knowledge gap and raises important issues for how we understand urban climate responses. It does so by drawing together insights from a wide range of countries, spanning the Global North to the Global South. The book is unique in its ambition and reach. It brings together 11 chapters by renowned urban climate governance scholars from around the globe. These chapters all critically assess the promises and limitations of increasing agency in urban climate governance. They build on solid empirical knowledge gained from case studies in the Global North and Global South. In doing so it sheds a much-needed critical new light on the existing literature and advances the current state of knowledge on urban climate policy and governance.

This book is part of the Earth System Governance (ESG) project, the world's largest social science research network of governance and climate change. Launched in 2009, ESG is a global research alliance connecting some 400 scholars who work on a range of themes, including urban climate governance.[1] In this book, we systematically cover the key research findings that have resulted from the project and related research activities. In this introductory chapter, we begin with setting out the key themes of the book – the politics of urban futures, increasing agency in urban climate policy and governance, and contested empowerment in urban transformations. We discuss some of the puzzles they raise for policy, practice, and academia, and propose a critical analysis of the heterogeneous forms of agency shaping the politics of urban futures. From here on, we briefly introduce the chapters that make up the main body of this book and how they relate to the broader ESG scholarship and other relevant communities and work in the field.

The Politics of Urban Climate Futures

Considering the urban politics and governance of climate change, much has happened since the 1990s (Bulkeley & Betsill, 2003; Jayne & Ward, 2017;

[1] See further www.earthsystemsgovernance.org (accessed 23 October 2017).

Parnell, 2016; Romero-Lankao et al., 2018). Three broad and related trends stand out that, as we will explain in what follows, shape the politics of urban climate futures: the combination of decentralization and liberalization, a growing ambition of local governments to bypass their national governments in urban climate policy and governance, and increasing recognition of the roles of local governments and local communities in global climate governance.

First, around the globe, two related developments have given local governments and local communities more influence in urban politics and the governance of local matters: decentralization and liberalization (Hodge, 2000; Taylor, 2013). Where local governments were once little more than the service-delivery branch of national governments and tasked with implementing national policies, they are increasingly expected to deliver local services themselves in an effective and efficient manner, and have to be transparent about their actions and be fully accountable for these – for instance through 'smart city' rankings and urban climate indexes (López-Ruiz, Alfaro-Navarro, & Nevado-Peña, 2014). In similar vein, the delivery of public services – waste collection, energy supply, and so on – is more and more tasked to the private sector and local communities through various forms of delegation, contracting out, and privatization (Osborne & Gaebler, 1992; Van der Heijden, 2011). This allows local governments and local communities to tailor urban policies to local needs, and use local resources for local service delivery.

Second, since the 1990s, local governments have been observed to adopt targets for reduced greenhouse gas emissions, energy consumption, and waste production well beyond the targets set by the nation states they are in (Acuto & Rayner, 2016; C40 Research Team & Arup, 2014; McKendry, 2018; Reckien et al., 2014). And since the 2000s, cities have become actively involved in climate adaptation also (Harman, Taylor, & Lane, 2015; Rauken, Mydske, & Winsvold, 2015; Rockefeller Foundation, 2013; Tanner et al., 2009). They are found actively involved in urban climate governance experiments that bring together local governments, private actors and civil society actors in formal and structured processes of developing, demonstrating and trialling new forms of authority and governance instruments to address climate challenges at the city level (Bulkeley, Castán Broto, & Edwards, 2015; Sassen, 2015). Of course, this is not to say that whenever they do so their main motivation is to reduce the effects of global climate change. Sometimes they are found to act simply to reduce the costs of operating cities, to prevent devastating effects of local climate-change–related disasters (Lovins, 2013; Nishida, Hua, & Okamoto, 2016), or hoping that an image of local climate action will attract investors and citizens that have a 'green' orientation and will ultimately boost the economic prosperity of their city (McCann, 2013; Schragger, 2016). Whatever their motivations the scale, scope and ambition of local government involvement in global climate governance has increased since, particularly, the mid-2000s (Bai,

Roberts, & Chen, 2010; Bulkeley, 2010). This becomes particularly evident when considering the emergence and growth of municipal networks at trans-local and international levels (Acuto & Rayner, 2016; Jayne & Ward, 2017; Jordan & Turnpenny, 2015; Lee, 2015). They allow cities to learn from each other, jointly experiment, and seek governance solutions to urban climate problems. Perhaps most importantly, they help local government to bypass their national governments in the international arena, and to raise awareness of their role in global climate governance.

Third and final, since the early 1990s there is increasing international recognition of the need to involve local governments and local communities in the politics of urban climate futures (Parnell, 2016). Already at the Earth Summit in Rio de Janeiro in 1992, local governments and communities were recognized and explicitly mentioned as an important site for climate action (UNCED, 1992). Following from this Earth Summit, international organizations such as the World Bank and the United Nations have led a range of initiatives to better understand the interactions between urbanization and climate change, with a particular focus on urbanization in the global south (UNDP, 2010, 2013; UNEP, 2007, 2011; World Bank, 2008, 2009a, 2011, 2013). Launched at the 21st Conference of the Parties (COP) in Paris, parallel climate summits for local and regional leaders have been held since that provide cities and other local actors with an opportunity to influence international climate change negotiations (Van der Heijden, 2018). Similarly, cities are a central focus of the United Nations' Sustainable Development Goals of 2015 (UN, 2015), and the New Urban Agenda resulting from the bi-decennial HABITAT Conference in 2016 has a strong focus on the role of cities in climate change mitigation and adaptation (United Nations, 2016).

Increasing Agency in Urban Climate Policy and Governance

It is against the backdrop of these three trends that novel agents of change and novel forms of agency engaged in governing climate change in the city have emerged. Agency is a contested concept and conceptualized differently across the social sciences (e.g. Alkire, 2008; Eisenhardt, 1989; Giddens, 1984; Sen, 1999). In a narrow understanding, agency reflects the capacity of individuals or organizations ('agents') to act independently and autonomously towards achieving desired outcomes. Social structures such as existing policy and governance arrangements for urban futures may help them achieving these goals, but are not considered to fully determine their behaviour, and may even be sources of deviance, improvisation, and entrepreneurship (Archer, 2003; Heugens & Lander, 2009). Such resistance against existing social structures may lead to changed social structures that increases the agency of some, but likely not others, which itself my lead to further resistance and, ultimately, change (Mahoney & Thelen, 2010; Van der Heijden & Kuhlmann, 2017). Of course, resistance may more likely result in change if agents are allowed to resist and have available social structures to resist (Parker, 2000; Stones, 2005).

In a broader understanding, agency is conceptualized as a propensity of social, socio-material, and socio-natural relations, and inhering not only in human individuals or organizations. Non-human agents (including animals, plants, and natural events) and more-than-human agency (including laws, technologies, procedures, and machines) may be considered having agency too, or, at the very least, affect the agency of humans (Gabriel, 2014; Latour, 2005; Murdoch, 2001; Swyngedouw, 2004). Floods, for example, may be considered having agency because they have 'the potential to be politically disruptive, with the ability, suitably mediated, to generate publics around it and to cast doubt on the status quo' (Donaldson et al., 2013: 611). Urban trees, on their turn, are sometimes considered to increasingly gain agency in the transition to sustainable cities, which replaces their traditional but passive aesthetic function in the urban landscape with an active and vital function in 'biogenic' or 'green' urban infrastructures (Kirkpatrick, Davison, & Daniels, 2013). Technological urban networks including water, waste collection, and communication technology, finally, may be considered to have agency as they empower those with access to it, and possibly strengthen inequalities between different groups of human agents (Leitner, Sheppard, & Sziarto, 2008).

The novel agents of change and novel forms of agency discussed in this book can be conceptualized, for simplicity, as either operating top-down, bottom-up, or mixing characteristics of both. Top-down approaches have, as highlighted before, a long history and can be traced back to the privatization of (local) public service delivery that started in the 1970s (Hodge, 2000; Van der Heijden, 2010), the 'reinventing' and decentralization of government and implementation of new public management practices since the 1980s (Hood, 1995; Osborne & Gaebler, 1992; Rondinelli, 1981), and the larger shift from government to governance that has been documented since the late 1990s (Rhodes, 1996, 2007). This all has tasked local governments with greater responsibility for achieving societally desired outcomes, including climate action, and has resulted in the involvement of a wide range of non-governmental agents in the regulation and governance of urban futures, including businesses, consultancies, universities, and research institutes (Bulkeley, 2013; Van der Heijden, 2014).

Bottom-up approaches for increased agency in urban climate policy and governance have been extensively discussed in, among others, grassroots literature (Seyfang & Haxeltine, 2012; Seyfang & Smith, 2007), civic engagement and social movement literatures (Brain, 2005; Portney & Berry, 2013), and self-governance and voluntary governance literatures (Acuto & Rayner, 2016; Kern & Bulkeley, 2009). It is in these literatures where we observe citizens and others to act, often purposefully, in ways that critically complement or even contrast existing urban policy and governance arrangements with an aim to ultimately change those arrangements. Typical examples are 'guerrilla gardening' and 'do-it-yourself urbanism' that make visible how small-scale interventions in urban public spaces may have long term benefits. Such interventions may initially be illicit, for example, setting up a vegetable garden on an unused plot of government land. Through phases of toleration, recognition, and enhancement the interventions may ultimately result in changed zoning laws that allow citizens to legally set up such gardens, but within certain regulatory boundaries (Finn, 2014; Hung, 2017). It is in these literatures also where we observe citizens, firms, and others to organize around a specific goal that is not yet included in urban climate policy and governance and voluntary commit to actions allowed within social structures that help to achieve it. A typical example is the global Transition Network movement, which brings together communities that seek to take, among others, climate action at local level, for instance a reduction of energy consumption through community energy projects (Brunetta & Baglione, 2013; Connors & McDonald, 2010; Smith, 2011). Another example is the organization of business interests around specific urban climate goals, for example, the emergence of Green Building Councils around the globe. These novel agents seek to support firms to increase their environmental sustainability behaviour or that of their products beyond government requirements, but do so on a paid-for basis (Van der Heijden, 2015).

Finally, mixed approaches have most clearly been captured in participatory governance literature (Brabham, 2009; Holden, 2011), collaborative governance literature (Clarke, 2016; Gollagher & Hartz-Karp, 2013), and network governance literature (Acuto & Rayner, 2016; Hughes, 2017). These literatures discuss a transition from traditional 'closed' decision-making processes towards those that involve citizens and other interested actors to different degrees – ranging from merely informing citizens, via consulting, to delegating power to citizens in urban policy and governance processes (Cheyne, 2015; Holden, 2011). By involving citizens and other relevant stakeholders in these processes their tacit knowledge can be used, diverse and competing views between agents might be bridged, and, ultimately, the outcomes of these processes may receive higher levels of acceptance (Bulkeley & Mol, 2003; Lobel, 2012). In the area of urban climate futures, these literatures are further aware of increasing collaborations between, particularly, local governments at regional, national and international level (Green, 2017; Lee, 2015). Over the years a number of these have formalized in powerful transnational municipal networks that now have a strong impact on the politics of urban climate futures, as discussed before. Three well-known networks are ICLEI – Local Governments for Sustainability (formerly named International Council for Local Environmental Initiatives), the C40 Cities Climate Leadership Group, and the Covenant of Mayors for Climate and Energy.

Contested Empowerment in Urban Transformations

Whether and to what extent has the increasing agency just described led to an increasing capacity for transformative urban climate governance? Has the overall capacity to transform urban futures in a low-carbon direction and address climate change increased also? Or, more critically, whether and to what extent has this increased agency empowered urban climate governance? Has increased agency helped to overcome some of the root causes of disempowerment? And how to gain insight and assess in this increased empowerment? As with the concept of agency, there is no shortage of definitions for and conceptualizations of the concept 'empowerment' within the social sciences (Bruce, 2007; Cornwall, 2010; Hur, 2006; Ibrahim & Alkire, 2007; Maru, 2009, 2011). It is by no means our ambition to provide a full overview of the discussions on empowerment, but a discussion of a few issues appears warranted considering the chapters that follow.

The notion of empowerment gained traction in the 1970s as an analytical construct to understand the development of individuals, organizations, and communities (Perkins & Zimmerman, 1995; Wilkinson, 1998). At its most basic level, empowerment can be understood as a redistribution of power (Hur, 2006; Wilkinson, 1998). Much scholarly work in this area studies the processes and means through which marginalized and oppressed individuals and collectives gain greater control over their lives, overcome barriers, resist existing power settings, emancipate, and achieve desired outcomes (Drydyk, 2013; Maton, 2008; Pease, 2002). Distinction is then often made between individual and collective empowerment (Hur, 2006; Moulaert et al., 2005; Rogers & Singhal, 2003; Rowlands, 1997). Individual empowerment concerns notions of self-determination and the capacity and competence to shape one's own life according to one's own desires, which includes being able to overcome social, institutional, and psychological obstacles (Drydyk, 2013; Hur, 2006). Collective empowerment concerns notions of mobilization, self-categorization, community building, and collective action aiming at social change beyond what individuals are able to achieve by themselves (Drury et al., 2005; Mohan & Stokke, 2000). Empowerment studies have traditionally been more concerned with individual empowerment than collective empowerment, but since the late 1990s the interest in collective empowerment has grown (Hur, 2006; Perkins & Zimmerman, 1995; Reininger et al., 2000). More recently, scholars have further opened up their analytical conceptualizations of and interest in empowerment and have begun to attribute empowerment to large collectives as well as non-human entities, including nature and technology (Kauffman & Martin, 2017; Sieber, 2006).

In studying empowerment, another distinction made is that between being 'empowered' and the 'empowering' processes, settings, actions and agency that may lead to empowerment (Hur, 2006; Maton, 2008) – a distinction between empowerment as an outcome and empowerment as a process (Drury et al., 2005; Drydyk, 2013; Mohan & Stokke, 2000; Moulaert et al., 2005; Reininger et al., 2000; Rogers & Singhal, 2003; Rowlands, 1997). There is a subtle difference between the two understandings: the former considers empowerment an end in itself (a virtue or norm), the latter considers empowerment a means to an end (Maru, 2009; Van der Heijden & Ten Heuvelhof, 2012). The former appears to have more resonance in studies on social welfare and social work, whereas the latter is more common in political science (Hur, 2006). Sherry Arnstein's ladder of citizen participation provides an appropriate example to illustrate this difference. As Arnstein suggests and others have empirically observed, sometimes, those in power involve citizens in decision-making processes, but only as a tokenistic gesture (Arnstein, 1969). They are not willing to truly share their power with them. Thus, citizens involved may de jure have been given more agency, de facto this will not help them improving the situation they are in. Those evaluating empowerment as an outcome would consider this a flawed or even failed form of empowerment (Arnstein, 1969; Oakley, 2001; Wilkinson, 1998). Those evaluating empowerment as a process may be milder in their judgement. That is, while the tokenistic form of participation is not helping citizens directly to improve the situation they find themselves in, it may show them how limited their power is and possibly unite them around that issue, which ultimately may raise resistance and calls for change. While tokenistic, this increased agency could then be considered a relevant empowering setting (Chavis & Wandersman, 1990; Hur, 2006).

A third and final for this book relevant distinction is that between what we, again for simplicity, conceptualize as legal empowerment and extra-legem empowerment (Drury et al., 2005; Drydyk, 2013; Golub, 2010; Hur, 2006; Ibrahim & Alkire, 2007; Mohan & Stokke, 2000; Moulaert et al., 2005; Reininger et al., 2000; Rogers & Singhal, 2003; Rowlands, 1997; UNDP, 2009). Legal empowerment is generally conceptualized as a process in which governments or other authorities provide legal or regulatory frameworks to empower particular disadvantaged groups and individuals; and, as a process in which individuals and groups use legal or regulatory frameworks to empower themselves (Bruce, 2007; Commission on Legal Empowerment of the Poor, 2008; Maru, 2009, 2011). These forms of empowerment are, in a sense, about the legitimacy of actors and the legitimacy of the agency given to them (Black, 2008; Haines, Reichman, & Scott, 2008). Extra-legem forms of empowerment are those in which individuals or collectives seek increased agency to govern and transform urban futures outside existing legal or regulatory frameworks. A well-known example is the various transnational municipal networks (TMNs) that have emerged since the 1990s. While they have little decision-making power at national or international levels, they have spurred and accelerated climate action at local level (Koon-hong Chan, 2016; Ljungkvist, 2016; Van der Heijden, 2018). Of course, the downside of extra-legem forms of empowerment is that approaches to govern and transform urban futures may be illegitimate or undemocratic, or result in illegitimate or undemocratic outcomes (Dawson, 2017; Hayden, 2014). For example, the 'climate proofing' of New York City with additional parks and green space was largely a result of wealthy residents organizing around this issue. Their neighbourhoods now indeed have increased green space, which has driven up property values. This drives out poorer residents from these neighbourhoods, only furthering social inequalities in the city (Gould & Lewis, 2016).

What follows from this discussion is that by no means can it be assumed that the increased agency discussed earlier leads to increasing capacity for transformative urban climate governance. Increased agency *in* urban climate governance is not synonymous with increased empowerment *of* urban climate governance. Increased agency may be tokenistic or distract from achieving a desired outcome, and inappropriate levels of agency may even be disempowering (Drury et al., 2005; Sieber, 2006). Finally, even when increased agency leads to increased empowerment of climate action, the outcomes of these actions may not be desirable. This book is therefore interested in which actors are becoming empowered to govern climate change, legitimately or otherwise, and what forms of agency are developed that enable and increase the capacity to address climate change.

When the Themes Meet: Puzzles for Policy, Practice, and Academia

These three themes – the politics of urban climate futures, increased and diversifying agency, and contested empowerment – raise important questions for urban climate policy, practice, and academia. For instance, what novel agents have emerged in urban climate governance since the early 1990s, and in what ways do they act? How is power given to or taken by them, and how do they exercise it? Who gains and who loses from the growing number of agents in urban climate governance? For example, the ever-growing number of agents in urban climate policy may have positive and negative impacts. An obvious positive impact is that the use of local agents to solve local urban climate problems may result in more adequate, tailored urban climate responses than traditional one size fits all urban climate policy solutions developed by somewhat distant agents. A possible negative impact may be that because of the high number of agents involved in urban climate policy and governance no one feels responsible for the overall outcome (cf., Thompson, 1980), or that sum of local urban climate policies will not add up to climate action that is adequate and timely at global level (Van der Heijden, 2017).

Another set of important questions revolves around novel forms of agency. Besides mapping the various forms of agency that have emerged in urban climate governance, we should ask which form of agency will yield the most desirable outcomes, and for whom. When considering, for example, the top-down emerged forms of agency discussed earlier, one may wonder whether this has truly resulted in increased and novel agency. It has been rightfully pointed out that these shifts have considerably increased the agency of some 'traditional' agents, for instance the important role mayors play in championing climate change plans and policies for their cities and at national and international platforms (Hughes, 2017). In yet other situations we might perhaps see a mere government orchestrated shift of agents (delegation of tasks from central government to local government, contracting out of tasks by local government to businesses, and so on) without a de facto change of agency (cf., Braithwaite, Coglianese, & Levi-Faur, 2007; Chhotray & Stoker, 2010).

In similar vein, policymakers, practitioners, and scholars face challenging questions on empowerment in urban climate transitions (Gramatikov & Porter, 2011; Svensson, 2006; UNDP, 2009; Van de Meene & Van Rooij, 2008). To what extent and how does including novel agents in urban climate governance empower them? To what extent and how does the combination of novel agents and novel agency empower climate action? Whether and how can (dis)empowerment in urban climate policy and governance be studied, made visible, and challenged? When, where and how can processes of empowerment be variable-sum (where everyone gains), and when, where and how are they zero-sum (where some gain and others lose)? For example, the purposeful clashing with existing policy and governance arrangements by guerrilla gardening and do-it-yourself urbanism is a different type of (self)empowerment than is the filling up of policy and governance voids and complementing existing arrangements as done by the Transition Network, or the de facto competition with these arrangements as is done by Green Building Councils. Is any of these types of empowerment more desirable than another, and on what grounds? And if so, to what extent are they transferable from one context to the next?

The most challenging puzzles are, however, the frictions between new agents of change, novel forms of agency, and the (dis)empowerment of individuals, organizations, communities, societies, and so on, to improve their well-being in urban climate futures. To name a few: Within the challenges and opportunities provided by urban climate futures, what are the most feasible forms of agency that have the largest empowering effect for climate action? To what extent and how can the capacity for transformation be improved without empowering individuals and collectives? To what extent and how can (and do) existing powerful agents purposefully increase agency without increasing empowerment?

Chapter Overview

With these and related questions in mind, the authors of this volume offer a multifaceted perspective on the emergence, current forms, and prospective trends of agency in urban climate governance. They explore the multiple and heterogeneous articulation of agency at different scales by unveiling the geographies of power supporting novel agents and new forms of agency, their innovative potential, and the resistance or resilience effects generated by the multiplication of agents of change in the urban context. The enclosed chapters present a coherent yet internally diverse set of contributions, addressing different scales, different geographies, different institutional structures, and different theories to explore and analyse agency and empowerment in the politics of urban future.

All in all, the theoretical breadth and depth of the chapters will be of interest to scholars who work on the edge of urban planning, policy science, public administration, and governance and are interested in the opportunities and constraints of different forms of agency in urban climate governance. The wide range of urban climate governance arenas, governance tools, and governance contexts illustrated with case studies and other examples and the way different actors seek to use and increase their agency in the politics of urban futures will be of interest to practitioners and policy makers in this field. Of course, as with any edited volume based on existing research, the extent to which the central themes of the book are given attention is limited by the material available. No novel research was carried out for this volume, but rather, theoretical and empirical linkages were sought across the work of ESG scholars to extract novel insights from existing studies. In similar vein, the geographical coverage of the book is limited also by existing work available. That having been said, case studies and real-world examples from cities around the world feature in the 11 chapters – including Accra and Kampala in Africa, Jakarta and Mumbai in Asia, Amsterdam and Berlin in Europe, Salvador and Rio de Janeiro in Latin America, and the Waterloo Region in North America.

This book includes two distinct sections. Chapters in the first section explore the geometries of power and responsibility in urban climate governance and their effect on agency and empowerment (Gordon; Patterson, & van der Grijp; Barbi & Valente de Macedo; Rickards; and Schulz & Bruns). Chapters in the second section focus on the empowerment processes in the urban context, and investigate the emergence and development of the multiple manifestations of engagement in the city space by individuals and organizations (Hughes et al.; Mancebo &

Certomà; Ninomiva & Burch; Uittenbroek et al.; and Gore; and Stehle et al.). Contributions in both sections range from the analysis of local-based strategies designed and adopted by groups of citizens in single cities, via the national level, to transnational networks of cities and other emerging forms of global agency. They provide insights from Europe, North and South America, Asia, and Africa, thereby offering a compelling panorama of agency and urban climate politics in cities and countries around the globe. Considering institutional structures, chapters are concerned with institutional initiatives led by local, national or supra-national governing or administrative bodies, including top-down, as well as bottom-up and non-formalized forms of citizen action. Chapters include theory-based considerations of social agency in the city development context and case-based analysis as illustrations of emerging agency in urban climate governance.

The first section opens with Gordon's contribution (Chapter 2) on the agency of transnational municipal networks whose very character is determined by the formation of collective identity, holding potential to overcome the limitations of localized action. Gordon highlights the way in which cities act to the call for global climate governance and rethink their role as global climate governor. This represents an important shift of cities from domestic contexts to the international arena, and important reconfigurations in terms of authority, legitimacy, and agency capability in world politics. Following from here, Patterson and van der Grijp's chapter (Chapter 3) uses a comparable systemic perspective on the analysis of transnational municipal networks and critically reflect on city empowerment and the role of administrations and peer networks in bringing about proactive behaviours towards political change. Through an analysis of climate change mitigation in Amsterdam, they illustrate that what cities can and cannot achieve is not only a function of 'empowerment' but also hinges on the broader socio-economic–political systems in which they are embedded.

The next contribution, by Barbi and Valente de Macedo (Chapter 4), addresses multilevel governance and the role of national and transnational municipal networks in the development and implementation of local climate strategies. They explore the contribution of the Council for Local Environmental Initiatives (ICLEI) and the Cities for Climate Protection (CCP) networks in supporting local government to experiment innovative intervention in sustainability strategies. In the chapter that follows, Rickards (Chapter 5) argues that within studies of climate change efforts, climates are recognized as urbanized or urban-made in three ways: global climate change, the deliberate production of certain climate-controlled spaces, and the direct but unintentional climate effects of the entire city landscape. Rickards points out that the latter issue remains neglected in the literature, and highlights the value of giving it more attention in future scholarship. The final contribution in the first part of this book, the chapter by Schulz and Bruns (Chapter 6), considers the internal relationships between different actors in their capacity to negotiate the future of urban climate governance. The authors discuss the role of new intersectional alliances in Accra and Berlin, drawing on urban political ecology theory. They show how both the cities underwent a long struggle over the governance of socio-technical systems which are crucial for climate change adaptation and mitigation.

The second section of the book opens with the contribution of Hughes et al. (Chapter 7) on data-driven governance. The authors point out the opportunities and risks of data-driven decision-making processes including disempowerment, access, and responsibility. They stress the need for cities to clearly identify the actors involved in the collection, storage, and use of data, and identify the most productive ways to navigate issues related to big data and smart city development for climate change mitigation. In the chapter that follows (Chapter 8), Mancebo and Certomà discuss possibilities to simultaneously fulfil sustainability and social justice priorities. As the authors highlight, this is a highly complex challenge in contemporary urban climate policy. They propose adaptive planning as an appropriate form of citizen agency to deal with the plethora of different issues emerging from the combination of social justice and sustainability. Using examples from urban agriculture projects they illustrate emerging forms of bottom-up and loosely coordinated agency, and explore the proactive role of citizens' participation in urban governance. The contribution by Ninomiya and Burch (Chapter 9) complements these insights. It is concerned with the nature of local climate action that extends beyond municipal governments. Using the example of the Decarbonize Waterloo Region initiative in Canada, they explore how actors new to the climate action milieu expand their agency and navigate the space of 'agency incumbents' (the utilities, elected officials and the public) at the local level.

From there on, Uittenbroek et al. (Chapter 10) explore climate adaptation strategies particularly as they relate to flooding problems in the Netherlands. The authors point out that there is little understanding of how to stimulate the development and implementation of policies to shift adaptation responsibilities from the public to the private agents (individuals and organizations). By analysing the City Deal on Climate Adaptation program in the Netherlands, they investigate how local governments perceive citizens' responsibilities in local climate adaptation, and what local governments are doing to induce citizens to take up these responsibilities. The question of how to stimulate private action in climate governance is also at the core of Gore's contribution (Chapter 11). Three cities in East Africa (Kampala, Dar es Salaam, and Nairobi) provide the context to explore the nexus between climate governance and food security. The author demonstrates that policy leadership may emerge in collaborative environments where urban civil society and government work together for the realization of innovative policy.

The final chapter, by Stehle et al. (Chapter 12), is interested in the effects of the activities of TMNs in a number of cities in Asia, Latin America, and Africa.

Finally, in the conclusion to the book we reflect on these contributions and synthesize the main arguments presented. This chapter presents the key lessons on agency and empowerment in urban climate politics and opens up challenging perspectives for a future research agenda on these themes in the politics of urban climate futures. It discusses the main agents found central to urban climate politics in the contributions to the book, the forms of agency applied by these agents, and seeks to find an answer to the question as to whether increased agency has led to increased empowerment in the politics of urban climate futures.

References

Acuto, M., & Rayner, S. (2016). City networks. *International Affairs*, 92(5): 1147–1166.

Alkire, S. (2008). Concepts and measures of agency. OPHI Working Paper, 9, 1–23.

Archer, M. S. (2003). *Structure, Agency, and the Internal Conversation*. Cambridge: Cambridge University Press.

Arnstein, S. A. (1969). A ladder of citizen participation. *JAIP*, 35(4): 216–224.

Bai, X., Roberts, B., & Chen, J. (2010). Urban sustainability experiments in Asia. *Environmental Science & Policy*, 13(4): 312–325.

Black, J. (2008). Constructing and contesting legitimacy and accountability in polycentric regulatory regimes. *Regulation & Governance*, 2(2): 137–164.

Brabham, D. (2009). Crowdsourcing the public participation process for planning projects. *Planning Theory*, 8(3): 242–262.

Brain, D. (2005). From good neighborhoods to sustainable cities. *International Regional Science Review*, 28(2): 217–238.

Braithwaite, J., Coglianese, C., & Levi-Faur, D. (2007). Can regulation and governance make a difference? *Regulation & Governance*, 1(1): 1–7.

Bruce, J. (2007). *Legal Empowerment of the Poor*. Washington, DC: U.S. Agency for International Development.

Brunetta, G., & Baglione, V. (2013). Resilience in the transition towns movement: Towards a new urban governance. *Journal of Land Use, Mobility and Environment*, 6(2): 251–264.

Bulkeley, H. (2010). Cities and the governing of climate change. *Annual Review of Environment and Resources*, 35(1): 229–253.

Bulkeley, H. (2013). *Cities and Climate Change*. New York: Routledge.

Bulkeley, H., & Betsill, M. (2003). *Cities and Climate Change*. Abingdon: Routledge.

Bulkeley, H., Castán Broto, V., & Edwards, G. (2015). *An Urban Politics of Climate Change*. Abingdon: Routledge.

Bulkeley, H., Castán Broto, V., & Gareth, E. (2012). Bringing climate change to the city. *Local Environment*, 17(5): 545–551.

Bulkeley, H., Edwards, G., & Fuller, S. (2014). Contesting climate justice in the city. *Global Environmental Change*, 25(March): 31–40.

Bulkeley, H., & Mol, A. (2003). Participation and environmental governance. *Environmental Values*, 12(2): 143–154.

C40 Research Team, & Arup. (2014). *Global Aggregation of City Climate Commitments*. New York: C40/Arup.

Castán Broto, V. (2017). Urban governance and the politics of climate change. *World Development*, 93(1): 1–15.
Chavis, D., & Wandersman, A. (1990). Sense of community in the urban environment. *American Journal of Community Psychology*, 18(1): 55–81.
Cheyne, C. (2015). Changing urban governance in New Zealand. *Urban Policy and Research*, 33(4): 416–432.
Chhotray, V., & Stoker, G. (2010). *Governance Theory and Practice*. Houndmills: Palgrave.
Clarke, S. (2016). Local place-based collaborative governance. *Urban Affairs Review*, 53 (3): 578–602.
Cohen, D. A. (2017). The other low-carbon protagonists. In M. Greenberg & P. Lewis (eds.), *The City Is the Factory*. Ithaca, NY: Cornell University Press,140–157.
Commission on Legal Empowerment of the Poor. (2008). *Making the Law Work for Everyone*. New York: United Nations.
Connors, P., & McDonald, P. (2010). Transitioning communities. *Community Development Journal*, 46(4): 558–572.
Cornwall, A. (2010). Introductory overview – buzzwords and fuzzwords. In A. Cornwall & D. Eade (eds.), *Deconstructing Development Discourse: Buzzwords and Fuzzwords*. Bourton on Dunsmore: Oxfam, 1–18.
Dawson, A. (2017). *Extreme Cities*. London: Verso.
Dodman, D. (2009). Blaming cities for climate change? *Environment and Urbanization*, 21 (1): 185–201.
Donaldson, A., Lane, S., Ward, N., & Whatmore, S. (2013). Overflowing with issues. *Environment and Policy C: Government and Policy*, 31(4): 603–618.
Drury, J., Cocking, C., Beale, J., Hanson, C., & Rapley, F. (2005). The phenomenology of empowerment in collective action. *British Journal of Social Psychology*, 44(3): 309–328.
Drydyk, J. (2013). Empowerment, agency, and power. *Journal of Global Ethics*, 9(3): 249–262.
Eisenhardt, K. (1989). Agency theory. *The Academy of Management Review*, 14(1): 57–74.
Finn, D. (2014). DIY urbanism. *Journal of Urbanism*, 7(4): 381–398.
Gabriel, N. (2014). Urban political ecology. *Geography Compass*, 8(1): 38–48.
Giddens, A. (1984). *The Constitution of Society*. Berkeley: University of California Press.
Gilbert, R., Stevenson, D., Girardet, H., & Stren, R. (2009). *Making Cities Work*. London: Earthscan.
Gollagher, M., & Hartz-Karp, J. (2013). The role of deliberative collaborative governance in achieving sustainable cities. *Sustainability*, 5(6): 2343–2366.
Golub, S. (2010). What is legal empowerment? In S. Golub (ed.), *Legal Empowerment: Practitioners' Perspectives*. Rome: International Development Law Organisation, 9–18.
Gould, K., & Lewis, T. (2016). *Green Gentrification*. London: Routledge.
Gramatikov, M., & Porter, R. (2011). Yes, I can: Subjective legal empowerment. *Georgetown Journal on Poverty Law & Policy*, 18(2): 169–200.
Green, J. F. (2017). The strength of weakness. *Climatic Change*, 144(1): 41–52.
Haines, F., Reichman, N., & Scott, C. (2008). Problematizing legitimacy and authority. *Law & Policy*, 30(1): 1–11.
Hall, P., & Pfeiffer, U. (2013). *Urban Future 21*. New York: E&FN Spon.
Harman, B., Taylor, B., & Lane, M. B. (2015). Urban partnerships and climate adaptation. *Current Opinion in Environmental Sustainability*, 12(1): 74–79.

Hayden, A. (2014). *When Green Growth Is Not Enough*. Quebec: McGill-Queen's University Press.
Heugens, P., & Lander, M. (2009). Structure! Agency! (And other quarrels). *Academy of Management Journal*, 32(1): 61–85.
Heynen, N. (2014). The urban century. *Progress in Human Geography*, 38(4): 598–604.
Hodge, G. A. (2000). *Privatization: An International Review of Performance*. Boulder, CO: Westview Press.
Holden, M. (2011). Public participation and local sustainability. *International Journal of Urban and Regional Research*, 35(2): 312–329.
Hood, C. (1995). The 'new public management' in the 1980s. *Accounting, Organizations and Society*, 20(2–3): 93–109.
Hughes, S. (2017). The politics of urban climate change policy. *Urban Affairs Review*, 53 (2): 362–380.
Hughes, S., Chu, E., & Mason, S., eds. (2018). *Climate Change in Cities*. Cham, Switzerland: Springer.
Hung, H. (2017). Formation of new property rights on government land through informal co-management. *Land Use Policy*, 63(April): 381–393.
Hur, M. H. (2006). Empowerment in terms of theoretical perspectives. *Journal of Community Psychology*, 34(5): 523–540.
Ibrahim, S., & Alkire, S. (2007). Agency and empowerment. *Oxford Development Studies*, 35(4): 379–403.
Jayne, M., & Ward, K., eds. (2017). *Urban Theory*. London: Routledge.
Jordan, A., & Turnpenny, J. (2015). *The Tools of Policy Formulation*. Cheltenham: Edward Elgar.
Kauffman, C., & Martin, P. (2017). Can rights of nature make development more sustainable? *World Development*, 92(April): 130–142.
Kern, K., & Bulkeley, H. (2009). Cities, Europeanization and multi-level governance. *Journal of Common Market Studies*, 47(2): 309–332.
Kirkpatrick, J., Davison, A., & Daniels, G. (2013). Sinners, scapegoats or fashion victims? *Geoforum*, 1: 165–176.
Koon-hong Chan, D. (2016). City diplomacy and 'glocal' governance. *Innovation: The European Journal of Social Science Research*, 29(2): 134–160.
Latour, B. (2005). *Reassenbkung the Social*. Oxford: Oxford University Press.
Lee, T. (2015). *Global Cities and Climate Change*. London: Routledge.
Leitner, H., Sheppard, E., & Sziarto, K. (2008). The spatialities of contentious politics. *Transactions of the Institute of British Geographers*, 33(2): 157–172.
Ljungkvist, K. (2016). *The Global City 2.0*. New York: Routledge.
Lobel, O. (2012). New governance as regulatory governance. In D. Levi-Faur (ed.), *The Oxford Handbook of Governance*. Oxford: Oxford University Press,65–82.
López-Ruiz, V.-R., Alfaro-Navarro, J.-L., & Nevado-Peña, D. (2014). Knowledge-city index construction. *Expert Systems with Applications*, 14(12): 5560–5572.
Lovins, A. (2013). *Reinventing Fire*. New York: Rocky Mountains Institute/Chelsea Green Publishing Co.
Mahoney, J., & Thelen, K. (2010). A theory of gradual institutional change. In J. Mahoney & K. Thelen (eds.), *Explaining Institutional Change*. New York: Cambridge University Press, 1–37.
Maru, V. (2009). *Access to Justice and Legal Empowerment*. Washington, DC: World Bank.
Maru, V. (2011). Allies unknown. *Health and Human Rights in Practice*, 12(1): 83–93.
Maton, K. (2008). Empowering community settings. *Community Psychology*, 41(1): 4–21.

McCann, E. (2013). Policy boosterism, policy mobilities, and the extrospective city. *Urban Geography*, 34(1): 5–29.

McKendry, C. 2018. *Greening Post-Industrial Cities*. New York: Routledge.

McKinsey Global Institute. (2011). *Urban World*. Seoul: McKinsey Global Institute.

Mitlin, D., & Satterthwaite, D. (2013). *Urban Poverty in the Global South*. London: Routledge.

Mohan, G., & Stokke, K. (2000). Participatory development and empowerment. *Third World Quarterly*, 21(2): 247–268.

Moulaert, F., Martinelli, F., Swyngedouw, E., & Gonzalez, S. (2005). Towards alternative model(s) of local innovation. *Urban Studies*, 42(11): 1969–1990.

Murdoch, J. (2001). Ecologising sociology. *Sociology*, 35(1): 111–133.

Nijkamp, P., & Kourtit, K. (2013). The 'New Urban Europe.' *European Planning Studies*, 21(3): 291–315.

Nishida, Y., Hua, Y., & Okamoto, N. (2016). Alternative building emission-reduction measure. *Building Research & Information*, 44(5–6): 644–659.

Oakley, P. (2001). *Evaluating Empowerment*. Oxford: INTRAC.

Osborne, D., & Gaebler, T. (1992). *Reinventing Government*. Reading, MA: Addison-Wesley Publishers.

Parker, J. (2000). *Structuration*. Buckingham: Open University.

Parnell, S. (2016). Defining a global urban development agenda. *World Development*, 78 (February): 529–540.

Pease, B. (2002). Rethinking empowerment. *The British Journal of Social Work*, 32(2): 135–147.

Perkins, D., & Zimmerman, M. (1995). Empowerment theory, research, and application. *American Journal of Community Psychology*, 23(5): 569–579.

Portney, K., & Berry, J. (2013). Civil society and sustainable cities. *Comparative Political Studies*, 47(3): 395–419.

Rauken, T., Mydske, P. K., & Winsvold, M. (2015). Mainstreaming climate change adaptation at the local level. *Local Environment*, 20(4): 408–423.

Reckien, D., Flacke, J., Dawson, R., et al. (2014). Climate change response in Europe. *Climatic Change*, 122(1): 331–340.

Reininger, B., Martin, D., Ross, M., Sminth Sinicrope, P., & Dinh-Zarr, T. (2000). Advancing the theory and measurement of collective empowerment. *International Quarterly of Community Health Education*, 19(4): 293–320.

Rhodes, R. A. W. (1996). The new governance. *Political Studies*, 44(4): 652–667.

Rhodes, R. A. W. (2007). Understanding governance. *Organization Studies*, 28(8): 1243–1264.

Rockefeller Foundation. (2013). 100 Resilient Cities. http://100resilientcities.rockefellerfoundation.org/ (accessed 7 January 2019).

Rogers, E. M., & Singhal, A. (2003). Empowerment and communication. *Annals of the International Communication Association*, 27(1): 67–85.

Romero-Lankao, Patricia, Sarah Burch, & Sara Hughes. 2018. Governance and Policy. In Cynthia Rosenzweig, William Solecki, Patricia Romero-Lankao, Shagun Mehrotra, Shobhakar Dhakal, & Solava Ibrahim (eds.), *Climate Change and Cities: Second Assessment Report of the Urban Climate Change Research Network*. Cambridge: Cambridge University Press, 585–606.

Rondinelli, D. (1981). Government decentralization in comparative perspective. *International Review of Administrative Sciences*, 47(2): 133–145.

Rowlands, J. (1997). *Questioning Empowerment*. Oxford: Oxfam.

Sassen, S. (2015). Bringing cities into the global climate framework. In C. Johnson, N. Toly, & H. Schroeder (eds.), *The Urban Climate Challenge*. London: Routledge, 24–36.

Schragger, R. (2016). *City Power*. Oxford: Oxford University Press.

Sen, A. (1999). *Development as Freedom*. Oxford: Oxford University Press.

Seyfang, G., & Haxeltine, A. (2012). Growing grassroots innovations. *Environment and Planning C: Government and Policy*, 30(3): 381–400.

Seyfang, G., & Smith, A. (2007). Grassroots innovations for sustainable development. *Environmental Politics*, 16(4): 584–603.

Sieber, R. (2006). Public participation geographic information systems. *Annals of the Association of American Geographers*, 96(3): 491–507.

Smith, A. (2011). The transition town network. *Social Movement Studies: Journal of Social, Cultural and Political Protest*, 10(1): 99–105.

Stones, R. (2005). *Structuration Theory*. Houndmills: Palgrave Macmillan.

Svensson, J. (2006). Bottom-up vs top-down approaches in peacebuilding and democratisation. In L. Rudebeck (ed.), *Violent Conflict and Democracy – Risks and Opportunities*. Uppsala: Uppsala University, 15–24.

Swyngedouw, E. (2004). *Social Power and the Urbanization of Water*. Oxford: Oxford University Press.

Tanner, T., Mitchell, T., Polack, E., & Guenther, B. (2009). *Urban Governance for Adaptation*. Brighton: Institute of Development Studies at the University of Sussex.

Taylor, P. (2013). *Extraordinary Cities*. Cheltenham: Edward Elgar.

Thompson, D. F. (1980). Moral responsibility of public officials. *The American Political Science Review*, 74(4): 905–916.

UN. (2016). *The World's Cities in 2016*. New York: United Nations.

UNCED. (1992). *Agenda 21*. Rio de Janeiro: United Nations.

UNDP. (2009). *Envisioning Empowerment*. New York: United Nations Development Programme.

UNDP. (2010). *Promoting Energy Efficiency in Buildings*. New York: United Nations Development Programme.

UNDP. (2013). *Addressing Urban Poverty, Inequality, and Vulnerability in a Warming World*. Bangkok: United Nations Development Programme.

UNEP. (2007). *Buildings and Climate Change*. Paris: United Nations Environment Programme.

UNEP. (2011). *Sustainability and Equity*. New York: United Nations Environment Programme.

United Nations. (2015). *Transforming Our World*. New York: United Nations.

United Nations. (2016). *New Urban Agenda*. Quito: United Nations General Assembly.

UN-HABITAT. (2016). *World Cities Report 2016*. Nairobi: UN-HABITAT.

Van de Meene, I., & Van Rooij, B. (2008). *Access to Justice and Legal Empowerment*. Leiden: Leiden University Press.

Van der Heijden, J. (2010). Smart privatization. *Journal of Comparative Policy Analysis*, 12 (5): 509–525.

Van der Heijden, J. (2011). Friends, enemies or strangers? *Law & Policy*, 33(3): 367–390.

Van der Heijden, J. (2014). *Governance for Urban Sustainability and Resilience*. Cheltenham: Edward Elgar.

Van der Heijden, J. (2015). On the potential of voluntary environmental programmes for the built environment. *Journal of Housing and the Built Environment*, 30(4): 553–567.

Van der Heijden, J. (2016). Opportunities and risks of the 'New Urban Governance' in India: To what extent can it help addressing pressing environmental problems? *Journal of Environment and Development*. DOI:10.1177/1070496516642500.
Van der Heijden, J. (2017). *Innovations in Urban Climate Governance*. Cambridge: Cambridge University Press.
Van der Heijden, J. (2018). *Cities and Sub-national Governance*. In A. Jordan & D. Huitema (eds.), *Policentricity in Action*. Oxford: Oxford University Press.
Van der Heijden, J., & Kuhlmann, J. (2017). Studying incremental institutional change. *Policy Studies Journal*, 45(3): 535–554.
Van der Heijden, J., & Ten Heuvelhof, E. (2012). The mechanics of virtue. *Environmental Policy and Governance*, 22(3): 177–188.
Watson, V. (2009). Seeing from the South. *Urban Studies*, 46(11): 2259–2275.
Wilkinson, A. (1998). Empowerment. *Personnel Review*, 27(1): 40–56.
World Bank. (2008). *Building Resilient Communities*. Washington, DC: World Bank.
World Bank. (2009a). *Climate Resilient Cities*. New York: World Bank.
World Bank. (2009b). *Systems of Cities*. New York: World Bank.
World Bank. (2011). *Supporting Efforts to Scale Capacity for Managing Urban Transformation*. New York: World Bank Institute.
World Bank. (2013). *Planning and Financing Low-Carbon, Livable Cities*. Washington, DC: World Bank.
World Bank Group. (2016). *Taking on Inequality*. Washington, DC: The World Bank.
Wright, E. O. (2010). *Envisioning Real Utopias*. London: Verso.

2

Unpacking Agency in Global Urban Climate Governance

City Networks as Actors, Agents, and Arenas

DAVID J. GORDON

Introduction

Cities, once relegated to the margins of the global response to climate change, have come to stake out for themselves a position at the front and center of the pack. Recent years have seen cities celebrated as global leaders and lauded for their propensity for pragmatic action and their capacity to drive transformative change (Barber, 2013). Importantly, city leadership is primarily associated with participation in transnational municipal networks (TMNs) such as the C40 Cities Climate Leadership Group, 100 Resilient Cities, and Global Compact of Mayors. These TMNs claim not only to facilitate urban policy innovation and experimentation, but more importantly also to produce coordinated action and meaningful collective effects.

In this chapter, I set out to assess TMNs through the lens of agency: what kind of agency do cities possess as global climate governors, where does this agency come from, and how has it changed over time? Whereas the individual agency of cities in global governance remains a source of ongoing debate, I draw out in this chapter the implications of understanding city agency as collective and global, as opposed to individual and localized, and suggest that doing so opens analysis up to an appreciation of the collective identity of cities as *global* climate governors.

To do so I begin with a brief overview of the empirical landscape and the conceptual terrain this chapter will traverse, with an emphasis on identifying the ways in which agency is defined and operationalized in the literature. From there the chapter pivots to a discussion of two distinct modes of city agency: the individual agency of cities and TMNs as sources of global climate governance and the collective agency of TMNs and networked cities. Whereas numerous tools exist with which to think about the former (agency as a function of capacity, authority, legitimacy), the latter has received much less attention. Drawing attention to its relationship to shared identity, I propose *recognition* as a key mechanism through which the constitution of collective urban agency takes place. I then illustrate the potential insights to be gained from this approach through a brief discussion of global accountability as a means through which collective urban agency has been produced. In brief, I suggest that being globally accountable serves to secure recognition (and access to highly sought resources) *for* cities as authoritative and legitimate global governors. At the same time, however, it creates a means of securing authority *over* cities by driving deference to a particular set of rules, processes, and practices – those related to counting, recounting, and accounting for governance actions and effects.

The intersection of these two processes – the constitution of an externally oriented collective agency and the crystallization of internally oriented authority relations – offers a potentially useful means through which to better understand the promise and perils of contemporary global urban governance. The chapter concludes by signaling several ways in which an expanded engagement with urban agency, of the sort set out herein, can contribute to better understanding and informing the role of cities in pursuing and achieving a just and equitable global sustainability transition (Bernstein & Hoffmann, 2018).

The Domain of Global Urban Climate Governance

Cities now have a three-decade-long history of engagement in the global governance of climate change. The manner in which they do so, however, has evolved over time. Bulkeley (2010) identifies an important shift from first to second "wave" of urban climate governance and associates it with an increase in the scale, scope, and ambition of city efforts (see also Acuto & Rayner, 2016; Arup, 2014, 2015; Aylett, 2014). Whereas early city efforts at climate governance were most often (a) untethered to practical plans, or capacities, for implementation (Betsill & Bulkeley, 2007: 448; Bulkeley & Kern, 2009; Krause, 2011) or (b) applied in a delimited manner to the operations or assets of municipal governments (Bulkeley & Betsill, 2003; Francis & Feiock, 2011), in recent years cities have come to link their nominal commitments to concrete practices and policy actions, albeit in more or less coherent or convincing ways. TMNs, in turn, now claim a position of global leadership in the response to complex issues of environmental degradation and societal transformation (Bloomberg, 2015; C40, 2016).

My intent in this chapter is to refract this phenomenon through the conceptual lens of agency. If we are to better understand the capacity of cities to be global leaders, there is a need to better understand the nature of their claims to act as global governors, as *agents* capable of producing (or at the very minimum creating the conditions for) systemic transformations of the sort required to achieve a meaningful global response to the challenge of climate change.[1] This, in turn, requires conceptual tools with which to understand (a) the nature and location of agency, (b) the foundations on which claims to agency rest, and (c) the processes through which such agency of cities is produced (and potentially contested). These questions, which fit into larger cross-disciplinary debates on the topic of agency, resonate strongly within the Earth System Governance (ESG) community. My goal here is both to highlight the substance of these contributions and to identify the limitations and lacunae that remain in need of further conceptual innovation and empirical analysis.

Agency and Global Urban Climate Governance

The fundamental distinction that I want to highlight is between scholarship that emphasizes the *individual agency* of cities and TMNs in global climate governance as opposed to that which highlights *the collective agency* of cities as global climate governors. As per the Merriam–Webster Dictionary, agency is "the capacity, condition, or state of acting or of exerting power."[2] It is, in the most basic sense, the capacity to act in, or upon, the world and signifies the ability to accomplish or realize objectives in a conscious and intentional manner. Agents, in other words, must be distinguished from actors; the latter are simply participants in the play of world politics whereas the former have the capacity to shape the plot and narrative itself. Accordingly, there is an important distinction between those whom we study as actors and those who possess meaningful agency in the global governance of climate change.

Broadly speaking, scholars of world politics have tended toward a state-centrism that prioritizes and locates agency in the state, although a strain of pluralist theory, which takes a more inclusive approach, is long-standing within the discipline (Schmidt, 2002) and has gained traction since the late 1980s. Yet while International Relations as a discipline has expanded its conceptual horizons to include consideration of the activities and effects of a diverse constellation of nonstate actors engaged in world politics (Avant et al., 2010; Bernstein & Cashore, 2012; Keck & Sikkink, 1998; Risse-Kappen, 1995), cities have remained largely absent from this scholarship, in part owing to their hybridity as actors neither fully private nor fully public, neither state nor nonstate (Bulkeley & Schroeder, 2011).

[1] Such efforts are visible across a variety of other issue areas, including security and counterterrorism, human migration and integration, communicable and noncommunicable health threats. See Ljungqvist (2016) and Acuto et al. (2017a, b).

[2] www.merriam-webster.com/dictionary/agency (accessed January 7, 2019).

The early work of Harriet Bulkeley and Michele Betsill, writing individually (Betsill, 2001; Bulkeley, 2005) and in collaboration (Betsill & Bulkeley, 2003, 2004, 2006), was central in redressing this limitation and paved the way for consideration of cities as meaningful participants in the global response to climate change. It also directed analytic attention toward the transnational networks through which cities linked their local activities to a broader global political project. This early work helped to spur the creation of a sub-field focused on cities and global climate governance that has grown substantially since the early 2000s. While it is a rich and diverse body of scholarship, for the purposes of this chapter I want to classify it into two camps and highlight the ways in which each reflects, and speaks to, the question of global urban agency.

Two Takes on Agency

Early scholarship on cities and global climate governance focused largely on establishing empirical and theoretical foundations on which to assess their potential contribution to the pursuit of global objectives. Inspired by the early efforts and engagement of cities – as manifest most clearly in the creation of TMNs such as the ICLEI – Local Governments for Sustainability (formerly named International Council for Local Environmental Initiatives) and the Climate Alliance in the early 1990s, and the adoption of greenhouse gas (GHG) emissions reduction targets by select cities around the world – this work focuses on the ways in which cities might stimulate meaningful change in urban settings and the efforts of TMNs to encourage, enhance, or augment their attempts to do so. This work pushed back on the notion that cities are perpetually subject to, or at the mercy of, the agency of others – whether that "other" be the state (Brutsch, 2013; Sancton, 2006), international organizations or multilateral agencies (Amen et al., 2011), private corporations or local interests (Molotch & Logan, 2007), or globally integrated corporations and advanced service providers (Sassen, 2001; Taylor, 2005). This work thus sought to locate, or at a minimum to explore the possibilities for, city agency, understood as the ability to pursue and produce meaningful effects on the world (Alber & Kern, 2008; Betsill & Bulkeley, 2003; Bulkeley et al., 2010; Kousky & Schneider, 2003). Cities, it has been convincingly argued, have access to (albeit to varying degrees) a set of levers with which they can actuate their individual agency as climate governors. They do so by encouraging and incentivizing citizens to act in particular ways, investing resources in providing services, establishing regulations and penalizing noncompliance, and demonstrating the benefits of action through control over their own city assets and employees (see Arup, 2016; Bulkeley & Kern, 2009; Bulkeley et al., 2010).

Importantly, the link between the local and the global in this line of scholarship rests largely on the participation of cities in transnational TMNs. These networks, from one point of view, serve to enhance the place-based agency of cities as global climate governors. This manner of thinking has a lineage that runs through the work of Margaret Keck and Kathryn Sikkink (1998) on transnational activist networks and the literature on epistemic communities (Haas, 1992), in the sense that TMNs are conceptualized as discrete organizational entities engaged in efforts to exert influence over some target audience. TMNs, for instance, work to activate and augment the individual agency of cities. They do so by sharing information and ideas, demonstrating the benefits of particular courses of action, establishing and incentivizing rules to guide member behavior, or facilitating access to material resources (Andonova et al., 2009; Bulkeley & Kern, 2009; Okereke et al., 2009; Selin & Vandeveer, 2009).

We can, as a result, think of the agency of the "network" – in terms of its ability to act as an entrepreneur, facilitator, or enabler – in encouraging the diffusion of norms, policies, and practices between cities through processes of learning and emulation (Hakelberg 2014; Lee & Koski, 2015; Lee & van de Meene, 2012; cf. Bouteligier, 2012). A primary takeaway is that the agency of "networks" to do so is both limited and highly contingent on structural (Lee, 2013); local institutional, political, and individual (Betsill & Bulkeley, 2003; Krause, 2011); and domestic (Gore, 2010; Bulkeley & Kern, 2009; Valente de Macedo et al., 2016) conditions. The agency of TMNs has further been found to be limited by their inability to compel compliance with nominal commitments, leading them to rely on alternative sources of authority (Bulkeley & Kern, 2009) underpinned by, for example, material resources, reputation, and organizational capacity (Gordon, 2016b). As a result, the agency of TMNs and cities is seriously constrained by the local conditions present in particular cities, including the presence of political leadership and policy champions, the institutional context, urban geography, and demographics (Aylett, 2014; Bulkeley & Betsill, 2003; Hughes, 2017).

Yet consider the following: in November of 2016 the C40 – *as a collectivity* – was presented with the annual Foreign Policy Green Diplomat of the Year award for its work on promoting climate action around the world.[3] This puzzling pronouncement opens up space to approach TMNs from a different perspective, as possessors of a *collective* global agency capable of producing autonomous governance effects. Whether or not this is indicative of a more fundamental process of transformation in world politics (Curtis, 2016; Sassen, 2006), it raises more

[3] www.c40.org/press_releases/press-release-foreign-policy-honours-c40-cities-climate-leadership-group-at-5th-annual-diplomat-of-the-year-awards (accessed April 29, 2018).

immediate and direct questions with respect the nature, foundations, and operation of global urban agency.

Acuto (2013a: 58) argues that the agency of cities in world politics can be understand as a function of their embedding in "multiscalar assemblages of global governance" (in the form of TMNs) rather than as discrete actors operating in their own right. Through an insightful application of actor–network theory, Acuto conceptualizes cities as possessing a collective agency, grounded on the presence of a collective *identity*. The collective agency of TMNs (or assemblages) thus has the potential to "supervene" the agency of their constituent members and become "capable of exerting influence on world affairs . . . " (Acuto, 2013b: 4).

Central to this story is the intriguing notion that agency is not located in the particular characteristics of the actors involved but rather is a product of socio-political processes through which they come together. Agency, in this telling, is relational, socially constructed, and contingent (on the broader debate regarding agency and authority in global governance see Krisch [2017] and Sending [2015]), and collective agency is closely tied to the constitution of shared identity. This assertion echoes the claim made by Erik Ringmar and Thomas Lindemann (2012: 3), who suggest that it is increasingly difficult to ignore identity if we are to understand the vagaries of world politics. With respect to cities, Kristin Ljungqvist (2016: 8) illustrates how this proposition can be leveraged to make sense of city participation in world politics, claiming that "[g]lobal [c]ity-hood as a specific type of collective identity can play a constitutive part in interest formation as local governments of certain cities claim political authority in foreign and security affairs and interact on their cities' behalf with the surrounding world on various global issue areas." Drawing tools from social psychology, Ljungqvist (2016: 26) argues that cities desire consciously to "be" a Global City (to achieve this status, in other words) and that this identity is not a singular but rather a *collective* one. Such "collective understandings of what it means to be a Global City" shape "a city's collective identity and interest formation," and this provides the foundation on which they assert a collective global agency in various domains. Importantly, the analytic focus rests on the question of consolidation of cities into collective actors – and the extent to which this provides an avenue for TMNs to overcome earlier constraints related to achieving coordinated action and collective effects.

Not only does this require that attention be paid to the efforts of states, and non–nation-state actors, to undertake "identity makeovers" through the telling of stories about themselves; it also demands sensitivity to the ways in which identity is narrated *to*, rather than solely *by*, actors. If identities are social facts created through social interaction (Ringmar & Lindemann, 2012: 5) then who tells these stories – whose stories shape the identity of the city as a global governor – are of utmost importance. For these stories serve to constitute cities as actors with agency of a particular sort.

The Constitution of Collective Urban Agency

In some sense this line of thinking circles back to an early interest in understanding "how ... governance take[s] place *within* transnational networks" (Betsill & Bulkeley, 2004: 476, my emphasis). A conceptual shift, however, is apparent, such that agency is no longer solely associated with the city or the "network" as a discrete actor, but instead is located in the network-as-collectivity. The network becomes a sociopolitical setting in which actors, drawing on differing claims to authority, seek to give shape and substance to the *collective* agency of the cities as global governors (Castán Broto, 2017). Central to this emphasis on identity and collective agency, then, is the proposition that agency is produced through socio-political relations between a variety of actors who occupy a shared sociopolitical space. In this section I want to offer some ideas as to how these might be addressed so as to generate a fuller understanding that highlights the politics and power dynamics inherent in contemporary global city climate governance.

A key question raised by recent scholarship is how cities can be (or are) "brought" together to forge a collective agency when their participation in TMNs is, by definition, voluntary and unenforceable (Chan & Pauw, 2014; Gordon & Johnson, 2017). A reconceptualization of agency as relational rather than dispositional – as produced through ongoing processes of social relations rather than being located within particular actors – provides a conceptual basis upon which to answer this question. It provides a means of interrogating why/how/when cities defer to the authority of others in terms of conforming to a particular set of governance ideas or practices, and how this constitutes a form of collective agency. In so doing it recognizes what Hickmann (2016) has identified as a "reconfiguration" rather than a zero-sum redistribution of authority in global climate governance, and provides a means of interrogating in greater detail the processes through which this reconfiguration has been taking place and is currently unfolding.

On this point Acuto (2013b: 7) proposes the importance of what he calls the "interessment" phase of network or assemblage formation, during which the identity and interests of participants are reshaped so as to conform to, or be rendered in accordance with, those of the assemblage as a whole. This is a powerful insight, but the resulting picture leaves questions open with respect to both how collective obligations or expectations are established and how they "work" to produce collective identity and pooled agency (through instrumental acquiescence, coercion, persuasion, and socialization). Thus, while Acuto (2013b: 16) suggests that " ... the [C40] has offered a chance for these metropolises to enhance their international legitimacy while also improving their policymaking independence" the causal link between this sort of instrumental calculus of cities and their adherence to a collective set of norms/expectations/practices remains unclear.

I propose recognition as a means of linking the individual members of a TMN and the constitution of a shared or collective identity (for an extended argument see Gordon [2015] and Gordon & Johnson [2017]). Brian Greenhill (2008: 344) argues that "recognition matters to international politics because it represents the process through which actors come to exist as actors within the international system and take on a particular identity within that system." This is an insight that can, I believe, be profitably turned toward the identity of cities as actors in world politics and the question of their collective agency as global climate governors. My suggestion is that recognition serves to create the conditions of possibility for the collective global agency of cities by leveraging the desire of cities for recognition from external audiences, and simultaneously using the promise of recognition to constitute and drive convergence around a common or shared identity. It is through recognition that relations of power become apparent.

Such a proposition does have some precedence – consider the assertion by Andonova et al. (2009: 56, my emphasis) that to be authoritative, or to possess agency, "[transnational] networks ... must be *recognized* as authoritative by the individuals and organizations that constitute the network." Rather than focusing on the status of *being* recognized (and directing attention toward the normative grounds on which such claims can and should be made), I suggest, instead, considering the merits of thinking about recognition as a means through which authority is claimed, agency is constituted, and collective identity is forged. It is, in other words, within the context of the TMN that an actor (or some subset of actors) claims the ability to secure recognition *for* cities. If deemed credible by others this then sets the terms on which that recognition is granted *to* cities. In this sense recognition opens the analytic door to considering the ways in which the constitution of collective city agency is suffused with power, albeit power operating outside the standard registers of command and compliance (Allen, 2010). Recognition is thus one way in which the power of cities (to achieve collectively meaningful effects) is fused to power over cities (required to bring order in the midst of diversity and absent the formal capacity to compel) (Partzsch, 2017).

Recognition underwrites claims to authority (as actors draw on various resources that they bring with them into a governance field) and motivates rule-following or norm adherence in conditions of nonhierarchical relations. It also represents a site of political contestation, as actors of various sorts (multinational corporations, nongovernmental organizations, international financial institutions, states, international organizations, philanthropic entities, and so on) with different ideas or interests compete to define what it takes to be recognized, by whom, and how recognition is to be secured. In so doing, an emphasis on recognition can help bring to light the political dynamics and power relations that operate beneath the surface of those efforts to mobilize information, facilitate learning, and disseminate best practices that are most often and explicitly associated with the activities of TMNs engaged in global climate governance.

Recognition and Global Accountability

To illustrate the analytic insights that might be gained from this approach I want to conclude with a brief vignette of climate governance in the C40 and the constitution of collective urban agency. The C40 has evolved, since its creation in 2005, from a loose collective to a well-funded and -staffed initiative; its membership has increased in number (from 40 to 92 cities as of early 2018) and diversity (more than half of C40 cities are located in the Global "South"); and C40 cities are observed to have increased the amount, scope, and ambition of their climate governance activities (Arup, 2015). In concert with these developments, the C40 has positioned itself – and cities more broadly – as an agent of innovation, experimentation, and global leadership (C40, 2014, 2016). The TMN, in other words, sees itself (and is seen by others) as endowed with *collective agency* in the sense that it can produce meaningful, intentional, and observable effects in/on the world.

There are indications that this collective agency is underpinned by the emergence of a specific collective identity. Cities of the C40 have come to converge around not only a common orientation with respect to the role of cities as crucial participants in the global response to climate change but also around particular governance practices and standards. These include, for example, uptake of the Global Greenhouse Gas Protocol for Cities (GPC), an emissions measurement and management standard developed jointly by C40, the World Resources Institute (WRI), and ICLEI, and practices of transparency and disclosure (of emissions inventories, governance activities, targets, and plans) to third-party platforms such as CDP Cities (Gordon, 2016a).

Together these constitute a means of simultaneously engendering and rendering legible the collective agency of the C40. Counting, recounting, and accounting for urban interventions provide a means with which the C40 can aggregate the collective commitments of its member cities and communicate these to global audiences such as the United Nations Framework Convention on Climate Change (UNFCCC) (C40, 2014). Indeed, beyond the C40 a good deal of effort and energy is being expended in an attempt to better quantify – and thus add up – city commitments, actions, and effects (Broekhoff et al., 2015; Erickson et al., 2014; UNEP, 2015). At the same time, these practices allow for the activities of cities to be compared both against each other and against themselves, through internal management systems such as the Arup/C40 Climate Actions in Megacities database and public-facing platforms such as the CDP Cities open dataset. In this way, the collective agency of the C40 is made both possible and apparent.

This is more than simply the diffusion of particular governance norms or policies, as cities in the C40 have come to adopt not a single norm (that of transparency, for example) or concrete project or policy (of emissions measurement or disclosure). Instead, these are indicative, I suggest, of a novel collective identity, that of the globally accountable city. The cities in the C40 have come to see themselves as participating in a shared project, and it has become more and more difficult for them to ignore the expectations that they embrace and they therefore enact their global or collective identity. On this point it is worth noting that this is not the first time that cities have toyed with a collective commitment to this combination of practices and norms. The ICLEI Cities for Climate Protection (CCP) TMN has been organized, since its inception in the early 1990s, around a five-milestone framework premised on an iterated process of setting quantifiable targets, measuring emissions, and tracking progress (Betsill & Bulkeley, 2003). This process, however, remained largely oriented toward enhancing the place-based agency of individual cities (helping them to demonstrate the economic benefits of policy intervention) and was never linked systematically to the production of a collective urban agency. The C40 itself engaged in a failed effort to establish a common emissions measurement and reporting methodology and platform (driven largely by its partnership with the Clinton Climate Initiative and in partnership with Microsoft) between 2007 and 2009 (Gordon, 2015). Why, then, has the C40 converged around this common identity now and not earlier?

Recognition offers a means of explaining why this is the case. Consider the following excerpt from a speech given by Christiana Figueras, then-head of the UNFCCC Secretariat, to the collected members of the C40 in early 2014. Figuer as suggested that to have a meaningful collective effect would require that cities

> First, use the metrics of the global conversation … set clean energy targets that are *recognizable* to the international process, and monitor progress towards meeting those targets … setting baselines and establishing inventories – with high quality accounting and reporting practices … [this] puts cities in the position to benefit from the various incentives and financial mechanisms that are being constructed both inside and outside the climate change convention. (Figueras, 2014, my emphasis)

This quote highlights, first, the extent to which the recognition of collective urban agency by global actors central to the climate regime is grounded on a specific set of expectations, ideas, and practices organized around novel systems of global accountability. By 2014 being globally accountable was, for cities, a means of becoming legible, of being *recognizable* to those audiences from whom they sought legitimacy and resources. That recognition is equated with measurement, standardization, monitoring, disclosure, and transparency. Yet these practices were not simply imposed upon the C40 from outside; indeed their contours were already apparent by 2011. In remarks made to a gathering of C40 cities, then-C40 Chair Michael Bloomberg explicitly set out a goal of securing recognition for the C40 as a leader in the global effort. Doing so would require cities to be "*100 percent accountable*" and to "set clear, *quantifiable benchmarks* for implementing them . . . [and] *regularly and openly assess our experience* with them" (Bloomberg, 2010 – my emphasis). An early effort at doing so took the form of a partnership forged in 2011 between the C40 and the World Bank. In return for securing from the World Bank the "landmark *recognition* of the leadership the world's great cities are taking to meet the challenges of climate change," the chair and leadership of the C40 undertook an effort to augment their ability to forge a collective agency (Bloomberg, 2011).

Why have the C40 cities largely conformed to these expectations, norms, and practices? One reason is that they provide a means of securing access to resources that enhance their local authority and augment their local governance capacity. Leveraging what Bloomberg referred to as the "tremendous opportunity" offered to "C40 cities to obtain vastly greater technical and financial support . . . [as well as] private capital" the partnership with the World Bank directly addressed the shared interest among cities in gaining access to public and private sources of financing and capital investment, but was made possible only by the "C40's commitment to standardizing how we report on the climate change plans in our cities" (Bloomberg, 2011). Similar dynamics are evident with respect to joint procurement initiatives such as the C40 Clean Bus Declaration (signed by 22 member cities) that identifies cumulative emissions reductions and commits signatories to integrating low-/no-emissions buses into public transit fleets by 2020.[4] The collective agency of the C40 is once again both constituted and rendered legible to global audiences – in this case market actors (manufacturers such as Volkswagen and BYD) and global funding agencies.

At the same time, accountability initiatives serve to disembed city officials from local contexts by rendering them visible to global audiences – augmenting peer pressure and enhancing access to reputational benefits. Accountability initiatives such as CDP Cities, standardized measurement and reporting tools, partnerships with funding sources such as the World Bank, and events such as the annual C40/Siemens Green Cities Awards (presented annually since 2013) all render C40 cities

[4] http://climateinitiativesplatform.org/index.php/C40_Clean_Bus_Declaration (accessed April 29, 2018).

increasingly recognizable not only to external audiences, but also to other network participants and partners. This offers cities a means of augmenting their capacity to compete for flows of transnational capital and investment[5] and shapes the terms on which cities are to be granted recognition from global audiences. In this sense, it provides a means through which the network is able to solidify its collective agency –by encouraging "'less active C40 members' and affiliate cities" to not only take up "more extensive actions" but also to drive convergence around particular sorts of norms, beliefs, and practices (Acuto, 2013b: 14).

Pathways Forward

In no small part facilitated by an explicit interdisciplinary orientation, ESG scholars have helped push scholarly debates beyond a narrow emphasis on the state, fostering an ontological pluralism in which cities are now seen as legitimate objects of analysis (Abbott, 2013; Bulkeley et al., 2014; Hoffmann, 2011). The collective agency of cities as global climate governors signifies a new dimension of the ongoing disembedding of cities from national/domestic contexts and a deeper process of reconfiguration with respect to the location and operation of authority and agency in world politics (Curtis, 2016; Sassen, 2006). In this concluding section I want to signal three important areas in which further research and creative thinking might be directed.

First, there is an opportunity to better assess and understand the tensions that accompany efforts to augment the collective agency of cities as global climate governors (Wachsmuth et al., 2016). As noted earlier, the scholarship on this topic highlights the myriad and diverse actors working to bring into form, and give focus to, the collective agency of cities but it remains unclear how these efforts map onto localized systems of democratic participation and representation – and how they redistribute, unsettle, or reproduce opportunities for participation and representation (Bäckstrand & Kuyper, 2017). Recent work on urban climate justice has begun to address these questions (Bulkeley et al., 2014, 2015; Van der Heijden, 2016), and there are interesting intersections with scholarship that looks at the relationship between domestic and transnational governance (Roger et al., 2017). There remains, nonetheless, considerable room for empirical and conceptual work focused explicitly on the intersection of global agency and domestic/local politics and the politics of inclusion and voice. Empirical research on adherence of cities to emerging collective identities, dynamics of recognition, and how these translate

[5] The recent call for proposals issued by Amazon for cities to host their secondary headquarters (colloquially referred to as HQ2) offers a potent illustration of this dynamic at play. See www.nytimes.com/2017/09/25/technology/wooing-amazon-second-headquarters.html?mcubz=1 (accessed April 29, 2018).

into local contexts is needed to better understand the implications of global urban agency.

A second line of inquiry relates to the need to better assess and understand the impact – both actual and potential – of global urban agency. Scholars have developed creative means of assessing the performance of cities, in terms of congruence between nominal commitment and policy action (Chan et al., 2017), the ambition of governance targets and policy actions (Bansard et al., 2016), and measurable effects of governance interventions (Erickson et al., 2014). Yet work is needed with respect to addressing whether cities can (and by whom) be held accountable for the promises and commitments (Gordon, 2016a; Hsu et al., 2016; Widerberg & Pattberg, 2017) and the empirical/conceptual markers through which agency can be detected and assessed (Gordon & Johnson, 2018; van der Ven et al., 2017).

Lastly, there is a need to think systematically about boundary conditions related to collective agency. Most research on global urban climate governance has focused on the activity of specific TMNs. Yet there are (as noted earlier) indications that cities as global governors are increasingly *networked* regardless of whether they participate actively or explicitly in specific TMNs. Acuto and colleagues (2017a; Acuto & Rayner, 2016) have made considerable headway in mapping the broader terrain of global urban governance, but there is work to be done on the interconnections between and beyond formal TMNs. Consider, as an illustration, the recent consolidation of the Global Compact and the EU Covenant of Mayors. The C40 sits at the core of this process, and questions have been raised as to whether it has gained prominence and authority vis-à-vis several other TMNs.[6] At the same time, it is worth noting that trends related to the uptake of practices of quantification, standardization, and disclosure noted earlier are observable well beyond the boundaries of the C40. How authority and power operate in this broader global urban domain, and with what implications for urban agency and the transformative impact of TMNs, are important questions to be addressed moving forward.

References

Abbott, Kenneth, W. (2013). Strengthening the transnational regime complex for climate change. *Transnational Environmental Law*, 3(1): 57–88.

Acuto, Michele. (2013a). *Global Cities, Governance and Diplomacy: The Urban Link*. London: Routledge.

Acuto, Michele. (2013b). The new climate leaders? *Review of International Studies*, 39(4): 835–857.

[6] www.euractiv.com/section/climate-environment/news/europeans-surrender-control-of-climate-initiative-to-bloomberg/ (accessed April 29, 2018).

Acuto, Michele, & Rayner, Steve. (2016). City networks: Breaking gridlocks or forging (new)lock-ins? *International Affairs*, 92(5): 1147–1166.

Acuto, Michele, Decramer, Hugo, Morissette, Mika, Doughty, Jack, & Ying, Yvonne. (2017a). *City New Frontiers for City Leaders. UCL City Leadership Lab Report*. London: University College London.

Acuto, Michele, Morissette, Mika, & Tsouros, Agis. (2017b). City diplomacy: Towards more strategic networking? Learning with WHO Healthy Cities. *Global Policy*, 8(1): 14–22.

Alber, Gotelind, & Kern, Kristine (2009). Governing climate change in cities: Modes of urban climate governance in multi-level systems. In *Competitive Cities and Climate Change, OECD Conference Proceedings*, Milan, Italy, October 9–10, 2008. Paris: Organisation for Economic Co-operation and Development, 171–196.

Allen, John. (2010). Powerful city networks: More than connections, less than dominance and control. *Urban Studies*, 47(13): 2895–2911.

Amen, Mark, Toly, Noah, McCarney, Patricia, & Segbers, Klaus, eds. (2011). *Cities and Global Governance: New Sites for International Relations*. London: Ashgate.

Andonova, Liliana, Betsill, Michele, & Bulkeley, Harriet. (2009). Transnational climate governance. *Global Environmental Politics*, 9(2): 52–73.

Arup. (2011). Climate action in megacities: C40 Cities baseline and opportunities. http://bit.ly/1dHnKn3 (accessed August 10, 2015).

Arup. (2014). Climate action in megacities: C40 Cities baseline and opportunities 2.0. http://bit.ly/1jfsIbJ (accessed April 29, 2018).

Arup. (2015). Climate Action in Megacities 3.0. www.cam3.c40.org/images/C40ClimateActionInMegacities3.pdf (accessed April 29, 2018).

Arup. (2016). Unlocking climate action in megacities. goo.gl/cYg8zP (accessed April 29, 2018).

Avant, Deborah, Finnemore, Martha, & Sell, Susan, eds. (2010). *Who Governs the Globe?* Cambridge: Cambridge University Press.

Aylett, Alex. (2014). *Progress and Challenges in the Urban Governance of Climate Change: Results of a Global Survey*. Cambridge, MA: MIT Press .

Bäckstrand, Karin, & Kuyper, Jonathan. (2017). The democratic legitimacy of orchestration: The UNFCCC, non-state actors, and transnational climate governance. *Environmental Politics*, 26(4): 764–788.

Bansard, Jennifer, Pattberg, Phillip H., & Widerberg, Oscar. (2016). Cities to the rescue? Assessing the performance of transnational municipal networks in global climate governance. *International Environmental Agreements*, 1–18.

Barber, Benjamin. (2013). *If Mayors Ruled the World: Dysfunctional Nations, Rising Cities*. New Haven, CT: Yale University Press.

Bernstein, Steven, & Cashore, Benjamin. (2012). Complex global governance and domestic policies: Four pathways of influence. *International Affairs*, 88(3): 377–403.

Bernstein, Steven, & Hoffmann, Matthew. (2018). The politics of decarbonization and the catalytic impact of subnational climate experiences. *Policy Sciences*, 51(2): 189–211.

Betsill, Michele. (2001). Mitigating climate change in U.S. cities: Opportunities and obstacles. *Local Environment*, 6: 393–406.

Betsill, Michele, & Bulkeley, Harriet. (2003). *Cities and Climate Change: Urban Sustainability and Global Environmental Governance*. New York: Routledge.

Betsill, Michele, & Bulkeley, Harriet. (2004). Transnational networks and global environmental governance: The Cities for Climate Protection Program. *International Studies Quarterly*, 48: 471–493.

Betsill, Michele, & Bulkeley, Harriet. (2006). Cities and the multilevel governance of global climate change. *Global Governance*, 12(2): 141–159.

Betsill, Michele, & Bulkeley, Harriet. (2007). Looking back and thinking ahead: A decade of cities and climate change research. *Local Environment*, 12(5): 447–456.

Betsill, Michele, & Bulkeley, Harriet. (2013). Revisiting the urban politics of climate change. *Environmental Politics*, 22(1): 136–154.

Bloomberg, Michael. (2010). Keynote Address at 2010 Hong Kong C40 Cities Workshop. http://on.nyc.gov/1Hay9ST (accessed April 29, 2018).

Bloomberg, Michael. (2011). Opening remarks by C40 Chair Michael Bloomberg at C40 Cities Climate Leadership Group Summit 2011. São Paulo, Brazil. http://bit.ly/1bBh1mK (accessed April 29, 2018).

Bloomberg, Michael. (2015). City century: Why municipalities are the key to fighting climate change. *Foreign Affairs* September/October 2015. www.foreignaffairs.com/articles/2015–08-18/city-century (accessed April 29, 2018).

Bouteligier, Sofie. (2012). *Cities, Networks, and Global Environmental Governance: Spaces of Innovation, Places of Leadership*. New York: Routledge.

Broekhoff, D., Erickson, P., & Lee, C. M. (2015). What cities do best: Piecing together an efficient global climate governance. SEI Working Paper 2015–15. www.sei-international.org/mediamanager/documents/Publications/Climate/SEI-WP-2015–15-Cities-vertical-climate-governance.pdf (accessed April 29, 2018).

Brutsch, Christian. (2013). From sovereign prerogatives to metropolitan rule? The anarchical society in the urban age. *International Studies Perspective*, 14: 307–324.

Bulkeley, Harriet. (2005). Reconfiguring environmental governance: Towards a politics of scales and networks. *Political Geography*, 24(8): 875–902.

Bulkeley, Harriet. (2010). Cities and the governing of climate change. *Annual Review of Environment and Resources*, 35: 229–253.

Bulkeley, Harriet. (2012). Governance and the geography of authority: Modalities of authorisation and the transnational governing of climate change. *Environment & Planning A*, 44: 2428–2444.

Bulkeley, Harriet, & Castán Broto, Vanessa. (2013). Government by experiment? Global cities and the governing of climate change. *Transactions of the Institute of British Geographers*, 38: 361–375.

Bulkeley, Harriet, Castán Broto, Vanessa, & Edwards, Gareth. (2015). *An Urban Politics of Climate Change: Experimentation and the Governing of Socio-Technical Transitions*. London: Routledge.

Bulkeley, Harriet, Edwards, Gareth, & Fuller, Sara. (2014). Contesting climate justice in the city: Examining politics and practice in urban climate change experiments. *Global Environmental Change*, 25: 31–40.

Bulkeley, Harriet, & Kern, Kristine. (2009). Cities, Europeanization and multi-level governance: Governing climate change through transnational municipal networks. *Journal of Common Market Studies*, 47(2): 309–332.

Bulkeley, Harriet, & Schroeder, Heike. (2011). Beyond state/non-state divides: Global cities and the governing of climate change. *European Journal of International Relations*, 18: 741–764.

Bulkeley, H., Andonova, L., Betsill, M., et al. (2014a). *Transnational Climate Change Governance*. Cambridge: Cambridge University Press.

Castán Broto, Vanessa. (2017). Urban governance and the politics of climate change. *World Development*, 93: 1–15.

C40. (2014). Working together: Global aggregation of city climate commitments. www.arup.com/publications/research/section/working-together-global-aggregation-of-city-climate-commitments (accessed April 29, 2018).

C40. (2016). Deadline 2020: How cities will get the job done. www.c40.org/researches/deadline-2020 (accessed April 29, 2018).

CDP. (2014). Protecting our capital: How climate adaptation in cities creates a resilient place for business. http://bit.ly/1UrINvz (accessed April 29, 2018).

Chan, Sander, & Pauw, Pieter. (2014). A global framework for climate action: Orchestrating non-state and subnational initiatives for more effective global climate governance, German Development Institute Discussion Paper 34/2014. http://bit.ly/2dJCNTw (accessed April 29, 2018).

Chan, S., Falkner, R. Goldberg, M., & van Asselt, H. (2017). Effective and geographically balanced? An output-based assessment of non-state climate actions. *Climate Policy*, 1–12.

Chan, S., van Asselt, H., Hale, T., et al. (2015). Reinvigorating international climate policy: A comprehensive framework for effective nonstate action. *Global Policy*, 6(4): 466–473.

Curtis, Simon. (2016). Cities and global governance: State failure or a new global order? *Millennium-Journal of International Studies*, 44(3): 455–477.

Erickson, P., & Tempest, K. (2014). Advancing climate ambition: How city-scale actions can contribute to global climate goals. SEI Working Paper No. 2014–06. Stockholm Environment Institute, Seattle, WA. http://sei-international.org/publications?pid=2582 (accessed April 29, 2018).

Figueres, Christiana. (2014). Keynote speech given at 2014 C40 Summit, Johanessburg, RSA (February 5, 2014). http://bit.ly/1isMyxg (accessed April 29, 2018).

Francis, N., & Feiock, R. C. (2011). *A Guide for Local Government Executives on Sustainable Energy Management*. Washington, DC: IBM Institute for the Business of Government.

Gordon, David J. (2013). Between local innovation and global impact: Cities, networks and the governance of climate change. *Canadian Foreign Policy Journal*, 19(3): 288–307. DOI:10.1080/11926422.2013.844186.

Gordon, David J. (2015). *From global cities to global governors: Power, politics and the convergence of urban climate governance*. PhD dissertation. University of Toronto.

Gordon, David J. (2016a). The politics of accountability in networked urban climate governance. *Global Environmental Politics*, 16(2): 82–100. DOI:10.1162/GLEP_a_00357.

Gordon, David J. (2016b). Lament for a network? Cities and networked climate governance in Canada. *Environment and Planning C: Government and Policy*, 20: 1–17.

Gordon, David J., & Acuto, Michele. (2015). If cities are the solution, what are the problems? The promise and perils of urban climate leadership. In C. Johnson, H. Schroeder, & N. Toly (eds.), *The Urban Climate Challenge: Rethinking the Role of Cities in the Global Climate Regime*. New York: Routledge, 63–81.

Gordon, David J., & Johnson, Craig. (2017). The orchestration of global urban climate governance: Conducting power in the post-Paris climate regime. *Environmental Politics*, 26(4): 694–714.

Gordon, David J., & Johnson, Craig. (2018). City-networks, global climate governance, and the road to 1.5°C. *Current Opinion on Environmental Sustainability*, 30: 35–41.

Gore, Christopher. (2010). The limits and opportunities of networks: Municipalities and Canadian climate change policy. *Review of Policy Research*, 27: 27–46.

Greenhill, Brian. (2008). Recognition and collective identity formation in international politics. *European Journal of International Relations*, 14(2): 343–368.

Haas, Peter. (1992). Epistemic communities and international policy coordination. *International Organization*, 46(1): 1–3.

Hakelberg, Lukas. (2014). Governance by diffusion: Transnational municipal networks and the spread of local climate strategies in Europe. *Global Environmental Politics*, 14(1): 107–129.

Hickmann, Thomas. (2016). *Rethinking Authority in Global Climate Governance: How Transnational Climate Initiatives Relate to the International Climate Regime*. London: Routledge.

Hoffmann, Matthew J. (2011). *Climate Governance at the Crossroads: Experimenting with a Global Response*. Oxford: Oxford University Press.

Hsu, A., Cheng, Y., Weinfurter, A., Xu, K., & Yick, C. (2016). Track climate pledges of cities and companies. *Nature*, 532: 303–306.

Hughes, Sara. (2017). The politics of urban climate change policy: Toward a research agenda. *Urban Affairs Review*, 53(2): 362–380.

Keck, Margaret, & Sikkink, Kathryn. (1998). *Activists beyond Borders: Advocacy Networks in International Politics*. Ithaca, NY: Cornell University Press.

Kousky Carolyon, & Schneider, Steven. (2003). Global climate policy: Will cities lead the way? *Climate Policy*, 3: 359–372.

Krause, Rachel M. (2011). Symbolic or substantive policy? Measuring the extent of local commitment to climate protection. *Environment and Planning C: Government and Policy*, 29(1): 46–62.

Krisch, Nico. (2017). Liquid authority in global governance. *International Theory*, 9(2): 237–260.

Lee, Taedong. (2013). Global cities and transnational climate change networks. *Global Environmental Politics*, 13(1): 108–127.

Lee, Taedong, & Koski, Chris. (2015). Multilevel governance and urban climate change mitigation. *Environment and Planning C: Government and Policy*, 33(6): 1501–1517.

Lee, Taedong, & van de Meene, Susan. (2012). Who teaches and who learns? Policy learning through the C40 Cities Climate Network. *Policy Sciences*, 45(3): 199–220.

Lindemann, T. (2010). The case for an empirical and social-psychological study of recognition in international relations. *International Theory*, 5(1): 150–155.

Ljungkvist, Kristin. (2016). *The Global City 2.0: An International Political Actor Beyond Economism?* London: Routledge.

Molotch, H., & Logan, J. (2007). *Urban Fortunes*, 2nd edn. Los Angeles: University of California Press.

Okereke, C., Bulkeley, H., & Schroeder, H. (2009). Conceptualizing climate governance beyond the international regime. *Global Environmental Politics*, 9(1): 58–78.

Partzsch, Lena. (2017). "Power with" and "power to" in environmental politics and the transition to sustainability. *Environmental Politics*, 26(2): 193–211.

Ringmar, E., & Lindemann, T., eds. (2012). *The International Politics of Recognition*. Boulder, CO: Paradigm.

Risse-Kappen, T. (1995). *Bringing Transnational Relations Back In: Non State Actors, Domestic Structures, and International Institutions*. New York: Cambridge University Press.

Roger, C., Hale, T., & Andonova, L. (2017). Domestic politics and climate governance. *International Interactions*, 43(1): 1–25.

Rosenau, James. (2003). *Distant Proximities: Dynamics beyond Globalization*. Princeton, NJ: Princeton University Press.

Ruggie, John. (1993). Territoriality and beyond: Problematizing modernity in international relations. *International Organization*, 47(1): 139–174.

Sancton, Andrew. (2006). Cities and climate change: Policy-takers, not policy-makers *Policy Options*, 27: 32–34.

Sassen, Saskia. (2001). *The Global City: New York, London, Tokyo*, 2nd edn. Princeton, NJ: Princeton University Press.

Sassen, Saskia. (2006). *Territory, Authority, Rights: From Medieval to Global Assemblages*. Princeton, NJ: Princeton University Press.

Schmidt, Brian. (2002). Anarchy, world politics and the birth of a discipline: American international relations, pluralist theory, and the myth of interwar idealism. *International Relations*, 16(1): 9–31.

Selin, Henrik, & VanDeveer, Stacy. (2009). *Changing Climates in North American Politics: Institutions, Policymaking and Multilevel Governance*. Cambridge, MA: MIT Press.

Sending, Ole Jacon. (2015). *The Politics of Expertise: Competing for Authority in Global Governance*. Ann Arbor: University of Michigan Press.

Taylor, Peter. (2005). Leading world cities: Empirical evaluations of urban nodes in multiple networks. *Urban Studies*, 42(9): 1593–1608.

Toly, Noah J. (2008). Transnational municipal networks in climate politics: From global governance to global politics. *Globalizations*, 5(3): 341–356.

UNEP. (2015). Climate commitments of subnational actors and business: A quantitative assessment of their emission reduction impact. United Nations Environment Programme (UNEP), Nairobi. http://uneplive.unep.org (accessed April 29, 2018).

Valente de Macedo, L., Setzer, J., & Rei, F. (2016). Transnational action fostering climate protection in the City of São Paulo and beyond. *disP – The Planning Review*, 52(2): 35–44.

Van der Heijden, Jeroen. (2016). Opportunities and risks of the "new urban governance" in India: To what extent can it help addressing pressing environmental problems? *Journal of Environment and Development*, 25(3): 251–275.

van der Ven, Hamish, Bernstein, Steven, & Hoffmann, Matthew. (2017). Valuing the contributions of nonstate and subnational actors to climate governance. *Global Environmental Politics*, 17(1): 1–20.

Wachsmuth, David, Cohen, Daniel Aldana, & Angelo, Hilary. (2016). Expand the frontiers of urban sustainability. *Nature*, 536: 391–393.

Widerberg, Oscar. (2017). The "black box" problem of orchestration: How to evaluate the performance of the Lima-Paris Action Agenda. *Environmental Politics*, 26(4): 715–737.

Widerberg, Oscar, & Pattberg, Philip. (2017). Accountability challenges in the transnational regime complex for climate change. *Review of Policy Research*, 34(1): 68–87.

Widerberg, Oscar, & Stripple, Johannes. (2016). The expanding field of cooperative initiatives for decarbonization: A review of five databases. *Wiley Interdisciplinary Reviews: Climate Change*, 7(4): 486–500.

3

Empowerment and Disempowerment of Urban Climate Governance Initiatives

An Exploratory Typology of Mechanisms

JAMES J. PATTERSON AND NICOLIEN VAN DER GRIJP

Introduction

Scholars have increasingly argued over the last decade that there are compelling opportunities as well as persistent challenges for climate action in cities, yet the overall implications for designing and pursuing urban climate initiatives remain unclear. Urban climate initiatives may take many different forms, such as policy innovation, experimentation, and urban laboratories (Evans & Karvonen, 2014) – all of which involve novel forms of agency seeking to influence urban governance systems to drive climate action. However, the existing literature on these topics remains piecemeal and fragmented from the perspective of informing strategic action. There is a key need to synthesize insights about ways in which empowerment/disempowerment of climate action in cities occurs, in order to understand the potential success or failure of future urban climate governance initiatives.

Urban climate governance initiatives may be empowered or disempowered by many different factors across different contexts. For example, this may relate to the presence of complex infrastructure systems, heterogeneous actors with contested interests, and intersecting structures of power and authority in urban governance (Aylett, 2013; Castán Broto, Oballa, & Junior, 2013; Hughes, 2017). Urban governance systems are often open-ended and not clearly demarcated from broader societal governance systems. Earth System Governance (ESG) scholars commonly view urban governance systems as multilevel (Betsill & Bulkeley, 2007), observing problems in divorcing the city from other jurisdictional levels (e.g. subnational, national, global) (Bulkeley & Betsill, 2005a: 43), and a frequent 'lack of "fit" between the nature of the problem to be governed and the institutions undertaking governance' (Betsill & Bulkeley, 2007; 450). Urban governance systems are also increasingly viewed as transnational (Bulkeley et al., 2014; Gordon & Johnson, 2017) owing to the emergence of a new urban climate change politics that blurs categories of authority and capability beyond those which can be easily captured by a traditional multilevel governance lens (Bulkeley & Betsill, 2013). In addition, urban governance systems are challenged by the differing scale of global problems (e.g. spatially, institutionally, temporally) contrasted against local capabilities to respond (Bai et al., 2010), leading some to argue for an 'open systems' view of cities (Bai et al., 2016). As a result, sources of agency for climate action are diverse, boundaries of urban climate governance systems are fuzzy, and factors causing empowerment or disempowerment of specific urban climate governance initiatives are multidimensional.

This chapter aims to critically review and synthesize the diverse ways in which urban climate governance initiatives may be empowered/disempowered. It develops an exploratory typology of mechanisms by which urban climate governance initiatives may be actively or passively empowered or disempowered. We consider urban climate governance initiatives as our unit of analysis. This reflects a collective (rather than individual) view of empowerment, where involved actors are linked through the common pursuit of an urban climate governance initiative. We focus on understanding the empowerment/disempowerment *of* an urban climate initiative within a complex urban governance system. The hypothesized logic of this approach is as follows: novel forms of agency (*causal conditions*) operate within structural contexts (*mediating conditions*) that may lead to shifts in power (*intermediate outcome*) and, ultimately, empowerment or disempowerment of the collective initiative over time (*outcome*). This causal logic differs from the way in which empowerment may be viewed from a vulnerability or social justice perspective, where the potential of urban climate governance to contribute to empowerment/disempowerment of certain social groups who have been marginal in conventional urban governance is scrutinized. In this chapter, we thus define empowerment as the process of enhancing the capability of a collective initiative to realize a desired outcome. Disempowerment is the inverse: the process by which the capability of a collective initiative to realize a desired goal is reduced.

The chapter first reviews existing literature based on the causal logic outlined in the preceding paragraphs, first considering forms of agency, structural conditions, and shifts in power, and then developing an exploratory typology of mechanisms of empowerment/disempowerment. We then present an illustrative case of empowerment/disempowerment in urban climate governance in Amsterdam, The Netherlands. Finally, we reflect on the lessons and next steps for scholars studying urban climate governance within the ESG network and beyond. The overall contribution of this chapter is a broad synthesis that can inform strategic action in urban climate governance, and lay a foundation for future studies that seek to understand successful urban climate governance particularly from an ex ante perspective.

Empowerment and Disempowerment in Urban Climate Governance

The theoretical approach in this chapter focuses on understanding how urban climate initiatives, which often begin on the margins of existing governance systems, become empowered or disempowered over time. This is important for understanding the potential success or failure of urban climate governance initiatives, as well as their impact on broader urban governance systems. Empowerment/disempowerment is a novel lens that goes beyond traditional capacity or enabler/barrier approaches, by bringing in a more dynamic and political perspective of processes of action and its effects in complex urban governance systems. The approach in this chapter defines empowerment/disempowerment analytically in relation to the interests of a group seeking to bring about a change in a governance system (i.e. an urban climate initiative seeking to promote greater climate action in a city). The approach is actor centred (and somewhat normatively agnostic) in that it can be applied to different initiatives working to different ends, which may not always align even within a single city. The approach is also 'action centred' in specifically seeking to understand how certain initiatives may be empowered/disempowered, which contrasts somewhat with a social justice or vulnerability perspective that typically focuses on understanding the empowerment/disempowerment of disenfranchised groups as an outcome of climate action.

The theoretical approach adopted involves ideas about agency, structure, and power. From a socio-economic development perspective, Alsop et al. (2005) developed a framework for studying empowerment centring on the dialectical relationship between agency and structure. They argued that looking at agency is not enough: it is also vital to examine the *opportunity structure* within which agency operates in order to understand whether desired change can in fact be realized. Following this approach, agency relates to the ability of actors seeking to take action (in our case, the actors involved in pursuing an urban climate governance initiative), and opportunity structure relates to the structural conditions within which action is pursued (in our case, the immediate urban governance system, as well as broader governance systems linked to it). Alsop et al. (2005) also pointed out that empowerment is a relational concept, and is closely linked to shifts in power between different actors. For example, Partzsch (2017) situated empowerment in the middle of a continuum of ways in which power may be mobilized in environmental governance: 'power with (cooperation and learning), power to (resistance and empowerment) and power over (coercion and manipulation)'. Therefore, understanding empowerment/disempowerment involves considering whether there have been shifts in power towards those actors seeking to take action. Importantly, this attention to potential shifts in power is what distinguishes an empowerment/disempowerment analytical lens from a conventional enabler/barrier analytical lens which typically treats power relationships as static and given.

This section briefly reviews urban climate governance literature relating to three key themes from the foregoing theoretical approach and articulated in the causal logic in the Introduction: (1) *Agency* (i.e. in what ways do actors involved in urban climate governance initiatives collectively exert novel forms of agency?); (2) *Opportunity structure* (i.e. what types of endogenous and exogenous structural factors influence the realization of urban climate governance initiatives?); and (3) *Shifts in power* (i.e. in which ways do urban climate governance initiatives gain or lose power in attempts towards their realization?). The next section then presents a typology of mechanisms of empowerment/disempowerment. Literature reviewed centres particularly on the ESG community, also bringing in related research by scholars with indirect links to this community. This provides a snapshot of emblematic issues and insights in urban climate governance from across diverse global regions.

Agency

Scholars have studied a variety of urban climate governance initiatives across the globe in recent years, identifying different ways in which actors may exert collective agency. This encompasses initiatives identified as policy innovation, experimentation, and urban laboratories (Evans & Karvonen, 2014).

Sources of collective agency for *policy innovation* largely centre on municipalities taking various initiatives. For example, in a study of urban climate governance in three Canadian cities, Burch (2010) identifies a multiplicity of concurrent initiatives including corporate strategic plans for municipal operations, municipal green building strategy, community planning, and interdepartmental collaboration. Scholars also highlight the importance of collective agency internally within municipalities which typically possess significant internal fragmentation. For example, in a study of urban climate governance initiatives in Durban and Portland, Oregon, USA, Aylett (2013) looked at the challenges of developing and integrating climate change strategies within the 'municipal bureaucracy', emphasizing the significance of internal heterogeneity (i.e. multiple sub-actors with different positions and interests). In a meta-analysis of urban climate change adaptation planning in the USA, Hughes (2015) also highlighted the importance of internal governance dynamics within a municipality such as between different departments (e.g. water, planning) and with elected representatives for successfully establishing urban adaptation planning initiatives. Looking outward, van Doren et al. (2016) conducted a detailed review of barriers to scaling up energy conservation initiatives for buildings, identifying various strategies that municipalities use to overcome barriers, which include informational (focusing on information and advice), cooperative (focusing on quality and efficiency of partnerships), financial (focusing on financial feasibility and attractiveness), and regulative (focusing on coercive measures to influence uptake) strategies.

Sources of agency for *experimentation* largely centre on non-state actors. As Bulkeley and Castán Broto (2014: 395) argue, focusing only on traditional forms of policy-making potentially leads to an 'impoverished picture of the challenges facing urban climate governance'. Starting from the premise that there is a crucial need 'to engage with the ways in which government is accomplished through social and technical practices', these scholars conduct a systematic global survey of urban experimentation, finding a rich landscape of activity occurring in non-traditional ways, where municipalities are central but do not operate alone (Bulkeley & Castán Broto, 2013). For example, many diverse actors may be key drivers of collective agency including private actors, community-based organizations, and non-governmental organizations, but these processes are also political and contested. For example, Bulkeley and Castán Broto (2014) conducted an in-depth case study in Banglore of a private initiative for low-carbon housing development driven by a private housing corporation seeking to promote green housing for low-carbon living to middle-class residents, revealing a complicated picture of innovation and equity-related implications. In the context of African cities' responses to climate change, Castán Broto et al. (2013) argued that diverse forms of agency including not only that of municipalities and private actors, but also of citizens, needs to be recognized as having an active role in urban climate change responses. These authors urged recognition of both planned and ad hoc forms of agency even in seemingly 'low-capacity' contexts of African cities.

Sources of agency for *urban laboratories* are understood to be inherently multi-actor. Urban laboratories for climate governance are commonly understood as 'spaces designed for interactions between a context and a research process to test, develop and/or apply social practices and/or technology to a building or infrastructure' (Voytenko et al., 2016). They aim to bring together key actors (e.g. municipalities, universities, private companies) to, in some way, formalize knowledge production about the real-world application of an urban technology or practice, ultimately geared towards upscaling of the novel initiative (Bulkeley & Castán Broto, 2013; Evans & Karvonen, 2014). Evans and Karvonen (2014) observe that urban laboratories, and scientific knowledge production more broadly, can be seen as an increasingly important source of agency in urban climate governance.

Opportunity Structure

The opportunity structure within which urban climate governance initiatives are pursued is here broadly separated into endogenous aspects (i.e. those lying within an urban climate governance system) and exogenous aspects (i.e. those lying beyond an urban climate governance system). This is a coarse distinction and oftentimes such separation is not neatly possible, but it serves as a useful starting point for organizing diverse factors. These factors closely reflect the longstanding focus of ESG research on multilevel governance (Bulkeley & Betsill, 2005b, 2013). The novelty of bringing these factors together from an empowerment perspective is that it allows them to be mobilized in new ways to inform strategic action (i.e. moving towards a diagnostic approach).

Endogenous Aspects

The *internal structure of municipalities* (e.g. within and between departments) is an important factor (Aylett, 2013; Bai et al., 2016; Bulkeley & Betsill, 2013; Hughes, 2017). For example, 'achieving coordination among sectors and departments can be a daunting task for local governments' (Hughes, 2017), but is of key importance and may often specifically be recognized in urban climate change plans (Hughes, 2015). Creating organizational structures that support collaboration and innovation is necessary (Burch, 2010). This is a particular challenge for cities in the Global South where such challenges compound with other capacity and resourcing gaps (Hardoy & Romero Lankao, 2011).

Organizational positioning of climate change matters; that is, where responsibility for addressing climate change is located institutionally. For example, whether it is situated at a high strategic level within a municipality and/or backed up by high-level mandates (Burch, 2010), or is allocated to a department with little actual power to push such initiatives (Hardoy & Romero Lankao, 2011). Also, whether climate change is viewed as an issue that competes or aligns with other concerns (e.g. sectoral issues, economic priorities) matters (Hughes, 2017). This may be related to motivations for addressing climate change in the first place, and whether a city is motivated more by internal or external factors (following Anguelovski & Carmin, 2011).

Organizational culture influences the realization of urban climate initiatives. For example, whether or not there is a collaborative and innovative culture is likely to matter greatly when initiatives cut across roles and responsibilities of different departments (Burch, 2010). Ideological tensions can arise due to different perspectives between different departments or between operational staff and planning officers, or as a result of narrowly focused leadership from senior managers (Aylett, 2013).

Previous climate change responses may affect the realization of further initiatives. For example, whether or not a municipality has prior experience with urban climate initiatives may affect overall willingness and momentum. Anguelovski and Carmin (2011) discussed the institutionalization of climate change responses, and how in the long-term the legitimacy and stability provided by institutionalization of adaptation initiatives is likely to be important for ongoing implementation and success. Van Doren et al. (2018) identified the importance of stable policy frameworks to foster trust in institutional arrangements among potential investors in low-carbon urban initiatives.

Availability of resources critically affects the realization of urban climate governance initiatives, even if strong political support exists (Hughes, 2017), for example, whether or not new revenue sources can be found if conventional sources are not enough (e.g. internal through property rates or other levies). Resources also matter for demand-side actors who are expected to invest in new initiatives (e.g. households, businesses) (van Doren et al., 2016).

More broadly, the *infrastructure context* involving social and technical practices as well as the materiality of infrastructure itself matters. These can be seen as comprising 'infrastructure networks' which fundamentally condition possibilities for urban experimentation (Bulkeley & Castán Broto, 2013, 2014). This also affects opportunities for scaling up initiatives (van Doren et al., 2016), for example, dealing with locked-in urban form (Burch, 2010).

The level of *awareness and support of users and wider citizens* can condition possibilities for realizing urban climate initiatives. For example, whether or not these groups provide bottom-up political support for bold action, or are aware of new opportunities that might aid in generating buy-in, may condition the possibility for effective scaling up of initiatives (van Doren et al., 2016).

Exogenous Aspects

Broader governance levels within which cities are embedded strongly condition response opportunities within cities, because often necessary political and legal authorities are dispersed across levels (Anguelovski & Carmin, 2011; Carmin, Anguelovski, & Roberts, 2012). As Hughes (2017: 369) states: 'The challenges of coordination extend beyond urban bureaucracies .... A broader set of actors – from the private sector, civil society, and other levels of government – participate in urban governance and shape outcomes in the city.' Hughes (2015) finds evidence of longstanding awareness among US municipalities of the importance of linking with higher levels of government in urban climate adaptation planning, but also challenges with maintaining such linkages over time. Burch (2010) charts diverse ways in which policy at provincial and national levels supported urban climate initiatives at the city level in several Canadian cities. This includes provincial planning and legal frameworks for climate change and energy, a carbon tax, emission standards and building codes, as well as general 'enthusiasm' that creates a supportive context locally.

Resourcing possibilities are critical in light of potential limitations at a local level. This may include possibilities for new funding streams from broader levels (e.g. investments, grants, allowance to create new levies). It is questionable whether exogenous sources should be relied on to fully close financing gaps on an ongoing basis. Yet urban climate initiatives are challenged by the long-term, open-ended nature of work often needed (Hughes, 2015). The interface with the market context matters for investment by private actors (e.g. capital costs, credit availability) (van Doren et al., 2016).

Broader political economic structures are directly relevant through impacting patterns of private investment linked to urban climate initiatives (e.g. linked to energy prices), as well as affecting the fiscal conditions of governments (e.g. domestic and global economies). More diffusely, these structures are connected to urban climate initiatives through the very expectation that cities take responsibility for dismantling deeply embedded systems of carbon-intensive production and consumption which are structural features of modern society.

The *socio-economic development context* is critical as there may be many pre-existing developmental and vulnerability-related challenges that urban climate initiatives must be sensitive to. For example, Hardoy and Romero-Lankao (2011) emphasize how Latin American cities 'are still faced with high levels of poverty, indigence and informality'. Hughes (2015: 17) observed that even in a North American context climate adaptation planning 'often lacks attention to equity issues, social vulnerability, and the influence of non-climatic factors on vulnerability'.

The role of various kinds of *crisis/shocks* is increasingly identified as a key factor influencing urban climate governance initiatives. This includes extreme weather events, but also crises/shocks of other forms. Aylett (2013) identified idiosyncratic interactions between various types of shocks which were conducive to urban climate initiatives, for example: (1) in Portland, Oregon, USA, a national financial crisis as well as public concern about domestic migration prompted by wildfires elsewhere in the country, and (2) in Durban, South Africa, a 'triple crisis' of energy shortages in the national electricity grid combined with food price increases linked to the global financial crisis and an extreme weather event which damaged infrastructure in the city. However, exactly under which conditions a crisis/shock is supportive or antagonistic to urban climate governance initiatives remains an open question.

Shifts in Power

It is hypothesized that urban climate governance initiatives gain or lose power through the interplay of agency and opportunity structure, as explained at the beginning of this section. Exactly how this plays out is likely to be subtle and highly case specific. Furthermore, shifts in power may sometimes be zero-sum, yet at other times they may be positive or negative sum. An example of a zero-sum shift is devolution, where new formal powers are conferred to municipalities from higher governance levels giving over authority (e.g. rule-making or revenue-raising capabilities). On the other hand, new forms of power may be created through bottom-up action within and among cities. For example, transnational municipal networks and private sector initiatives relating to cities (e.g. city rankings, sustainability/energy certification) create new patterns in roles and authorities with at least rhetorical or persuasive powers even if not formal.

More broadly, shifts in broader discourses (e.g. about the urgency of climate change, and the role of cities as global players) may open up new opportunities for cities to 'claim' new roles and authorities. However, there may also be practical limits on the creation of new formal powers at the city level, such as pushback by powerful political interests or constitutional limits on political authority. Understanding how the interplay between agency and opportunity structure leads to possible shifts in power, with consequences for the production of empowerment/disempowerment, is the topic of the next section.

Mechanisms of Empowerment and Disempowerment

This section considers how the three variable categories in the previous section (i.e. agency, opportunity structure, shifts in power) interact to produce empowerment/disempowerment. The logic here is that *agency* and *opportunity structure* interact and lead to possible *shifts in power*, which leads to *empowerment/disempowerment* outcomes. Shifts in power may or may not occur; that is, this variable may take two possible outcomes: shift or no shift. This may be associated with two possible outcomes: empowerment or disempowerment. This can lead to four different scenarios: active or passive empowerment, and active or passive disempowerment (Table 3.1).

Within the framework of Table 3.1, we synthesize a variety of possible mechanisms of empowerment and disempowerment that might occur across different contexts. Mechanisms are considered propositions about causal relations, which make the thinking behind causal claims explicit and testable (Beach & Pedersen, 2016). Formulating mechanisms of empowerment/disempowerment helps to make sense of disparate literature in a way that allows empirical testing, further

Table 3.1 *Production of empowerment/disempowerment*

	Empowerment	Disempowerment
Active	Urban climate initiative gains power through shifts from other actors or creation of new sources of power.	Urban climate initiative loses power through shifts to other actors or undermining of sources of power.
Passive	Urban climate initiative gains power through changes in the broader urban governance context.	Urban climate initiative loses power through changes in the broader urban governance context.

theorizing, and potentially also informs strategic action. This is timely because the field of urban climate governance has been in a relatively exploratory phase for the last decade or so, both theoretically and empirically, and arguably now needs to move into a consolidation phase to synthesize insights gained. Identification of common mechanisms across diverse contexts supports this goal.

Table 3.2 shows possible mechanisms of empowerment/disempowerment based on the reviewed literature. This aims to capture important dynamics that produce empowerment or disempowerment of urban climate governance initiatives. These dynamics may arise in ways that are endogenous or exogenous to a city. Whether or not a shift in power occurs may not always be unambiguous. The basic test we suggest is whether other actors in the system (who are not the subject) would recognize a shift in power or not. Many of these mechanisms are 'opposites' in the sense that they can contribute to empowerment or disempowerment depending on their directionality. The overall set of mechanisms listed is not exhaustive, but is emblematic of those that tend to be frequently observed across diverse contexts and thus potentially have widespread resonance. This helps to lay a foundation for future scholarship to analyse processes by which empowerment/disempowerment occurs, in a way that can be contextualized, but also retain a level of generality to enable cross-case comparison and synthesis.

Illustrative Case: Initiatives for Energy Transition in Amsterdam, the Netherlands

Overview of Initiatives

The specific issue that this case involves is energy provision in a medium- to large-sized city within a developed country context. Amsterdam has a population of approximately 850,000 within the city proper, and up to 2.4 m within the broader metropolitan area. Since the early 1990s, the City of Amsterdam has pursued climate mitigation and energy policies, but overall left much of the initiative to

Table 3.2 *Possible mechanisms of empowerment/disempowerment*

	Empowerment	Disempowerment
Active	Endogenous: • Senior city leadership encourages climate action, which confers mandate to urban climate initiative. • Strategic coalition of actors mobilizes in support of urban climate initiative. • Cultivation of narratives of climate action which discursively supports urban climate initiative.	Endogenous: • Senior city leadership discourages climate action, which undermines mandate for urban climate initiative. • Strategic coalition of actors mobilizes against urban climate initiative. • Cultivation of narratives that conflict with climate action which discursively undermines urban climate initiative.
	Exogenous: • Policy/legal frameworks created at higher levels that support climate action in cities. • Resources for urban climate action provided from higher levels. • City participates in transnational networks/initiatives that create new imperatives for climate action in cities.	Exogenous: • Policy/legal frameworks at higher levels constrain climate action in cities. • Resources for urban climate action withheld from higher levels. • Elite intervention from higher levels to block urban climate initiatives.
Passive	Endogenous: • Institutional voids provide space that is claimed by new urban initiatives. • Institutional dynamics create windows of opportunity to secure support (e.g. planning and political cycles). • Citizen awareness and support for climate action creates political support for urban climate action.	Endogenous: • Institutional voids create difficulty for new urban initiatives to gain traction. • Institutional dynamics make it difficult to secure or sustain support (e.g. locked-in plans and political agendas). • Lack of citizen awareness and support makes it difficult to build political support for urban climate action.
	Exogenous: • External crises/shocks generate broad support for climate action. • Municipal leadership participates in higher-level policy/planning forums. • Increasing prominence of transnational networks/initiatives that create new imperatives for climate action in cities.	Exogenous: • External crises/shocks reduce broad support (e.g. narrow recovery focus). • Apathy in higher-level policy arenas towards municipal involvement. • National/global economic conditions hinder public and private investment in climate action.

private actors, such as energy companies and local energy cooperatives (Hisschemöller, 2016). As a consequence, some positive results have been achieved but less overall progress than envisioned. However, around 2010 the city began to increase its engagement, publishing several policy documents aiming to provide a stronger impetus for change (van der Hoek, Struker, & de Danschutter, 2017). This led to investment in novel infrastructure including a district heating system utilizing surplus heat from municipal waste incineration, and the creation of a charging infrastructure for electric vehicles. However, overall these approaches were largely technocratic with little attention to citizen engagement.

Recently a renewed set of climate ambitions were set out by the municipality, formulating a strategy to phase out the use of natural gas for heating called the Amsterdam City Deal (City of Amsterdam, 2016). This involves a voluntary agreement between major stakeholders in energy and social housing, stipulating targets and actions for these participants, and involves business case development, pilot projects, and citizen engagement. It appears to be a significant step towards realizing an energy transition, marking a shift in climate policy in the city.

In parallel, citizens have become increasingly engaged in urban climate initiatives in the city, supported by environmental non-governmental organizations (NGOs) such as *Urgenda* and *Milieudefensie*. Around a dozen bottom-up citizen energy cooperatives are active in Amsterdam, stimulated by national economic incentives (such as the *Postcoderoosregeling* and *SDE+*). They initially focused on wind energy projects but subsequently became involved in solar energy and energy saving initiatives, and more recently district heating. Increasingly, these initiatives are joining forces, seeking to work together with the municipality, energy network operators, and private companies. This citizen energy movement and its partners recently began collaborating under the umbrella of 'Platform 02025' aiming for an ambitious urban energy transition by 2025 when the city celebrates its 750th birthday.

Recently the municipality has also cultivated a strong emphasis on knowledge coproduction through various projects and platforms. For example, the municipality is successful in acquiring European project subsidies, such as the City-zen project and the NEXT-Buildings project. It also has a well-developed cooperation with a broad range of stakeholders through its 'Smart City' initiative focusing on urban innovation, and with knowledge institutions such as the Amsterdam Institute for Advanced Metropolitan Solutions.

Empowerment/Disempowerment

Agency is being exerted by a range of actors towards policy innovation (e.g. Amsterdam City Deal), experimentation (e.g. citizen energy movement), and urban laboratories (e.g. knowledge coproduction initiatives).

The opportunity structure is conducive to these initiatives in various ways. Endogenously, the municipality has cultivated and responded to a city-wide agenda for energy system transition, and increasingly also frames its approach as part of a broader transition towards a 'circular city' in order to enhance its profile and appeal (Circle Economy et al., 2018; TNO et al., 2015). Outwardly, the municipality increasingly performs a strong role as an initiator and facilitator of climate action. Inwardly, it addresses the complex organizational structure of the municipality through the appointment of an alderman (councillor) for sustainability who is also responsible for climate policy. The city increasingly uses its strong legal position in land ownership as a catalyst for change (Savini et al., 2016). Almost the totality of urban land is owned by the city and leased to its users. This not only provides a stable source of income to the city but also allows the local government to directly control land use in order to pursue and implement municipal policies. Examples are the menu of sustainable building options from which project developers and real estate agents are asked to make a choice. Recently, the municipality issued a roadmap for the circular tendering of land which is meant as an instrument to stimulate, measure, and reward circular building and renovation, and which features energy issues prominently (Roemers & Faes, 2017). Exogenously, there is discursive alignment in agendas across municipal, provincial, and national governments in regards to an ambition for a transition towards sustainable energy systems ultimately by 2050. A major shared incentive is the rapid depletion of the finite Dutch natural gas resources as well as the increased occurrence of earthquakes in the northern part of the country, due to gas extraction. This has spillover effects on energy debates within Amsterdam among citizens, government, and industry; building awareness; and support of users and citizens.

Various active empowerment mechanisms are observed, including senior city leadership conferring a mandate for urban climate governance initiatives, strategic coalition of actors mobilizing in support of urban climate governance initiatives, cultivation of narratives of climate action which discursively support urban climate initiatives, and policy/legal frameworks created at higher levels that support climate action at the city level. At the same time, various passive empowerment mechanisms are also observed, including institutional voids providing space that is claimed by new urban initiatives, institutional dynamics creating windows of opportunity to secure support, and citizen awareness and support for climate action creating political support for urban climate action.

On the other hand, there is a risk of active disempowerment due to resources for urban climate action being withheld from higher levels. Financing for the energy transition outlined in the City Deal is estimated at 5 to 6 billion euro, yet sources of funding are yet to be fully identified. Thus far, the national government has not pledged any direct financial support to help realize the climate ambitions of

Amsterdam. This means that at the executive level, the municipality is largely dependent on other parties such as project developers, businesses, and citizens to achieve concrete results. It has itself only a limited budget for such financing. If resourcing is not made available by national and/or provincial governments, this may reflect a form of active disempowerment by withholding resources at important moments, leading to missed opportunities. At the same time, there is a risk of passive disempowerment due to institutional dynamics that may make it difficult to secure or sustain support because of other simultaneous plans and political agendas. For example, Amsterdam has high ambitions for urban growth (50,000 new dwellings ultimately in 2025), and elaborate administrative procedures for city development as established in the so-called 'Plaberum' (Marselis & Hisschemöller, 2018).

Altogether, this exploratory case study analysis illustrates how an analysis of empowerment/disempowerment of urban climate governance initiatives, as outlined previously, can be conducted. Interestingly, it reveals multiple empowerment/disempowerment mechanisms operating simultaneously. This places a novel forward-looking emphasis on dynamics unfolding, and opportunities for strategic action to improve the prospects of success of these initiatives towards urban climate action.

Discussion and Conclusions

Understanding Empowerment/Disempowerment

The key contribution of this chapter is a systematic approach, including an exploratory typology, for analysing empowerment/disempowerment of urban climate governance initiatives. This is important because urban climate governance initiatives are pursued within a multidimensional web of linkages and interdependencies, including aspects that are both endogenous and exogenous to an urban climate governance system. Empowerment or disempowerment is shaped not only by agency (i.e. through urban climate initiatives), but also by the opportunity structure within which such initiatives are embedded. Mechanisms identified provide a starting point for contextualized investigation in any particular case. These mechanisms may need to be further adapted or unpacked in a specific context, but they provide a starting point that is likely to have general resonance. An empowerment/disempowerment lens helps to understand the performance of urban climate initiatives, as well as their effects on urban governance systems. Yet it does not fully explain the emergence of urban climate initiatives in the first place, which may arise from diverse motivations not covered here.

Multiple mechanisms acting in combination may be necessary for producing empowerment. For example, in Bangalore Bulkeley and Castán Broto (2014) found that two simultaneous mechanisms were important for the initiative they studied: the cultivation of discourses reinforcing the logic of the initiative, and the presence of a strategic coalition of actors mobilizing powerful interests to support the initiative. This was also evident in the illustrative case of Amsterdam, where multiple mechanisms contributing to empowerment were identified (albeit with risks of disempowerment looking forward). Yet the interaction of multiple mechanisms may also produce disempowerment. For example, Hardoy and Romero-Lankao (2011) identify the simultaneous occurrence of non-provision of resources, obstructive institutional dynamics, and lack of ability to address wider socio-economic developmental issues for urban climate governance in Latin America.

On the other hand, Anguelovski and Carmin (2011) indicate that empowerment and disempowerment mechanisms may coexist (e.g. non-provision of resources and institutional voids, alongside policy entrepreneurship), with mixed implications for realizing urban climate governance initiatives. Furthermore, there may be temporal patterns in ways that empowerment/disempowerment is produced. Carmin et al. (2012) argue that 'early adoption' may be explained more through endogenous factors owing to a lack of existing exogenous incentives, whereas 'late adoption' may be explained more by exogenous factors due to the intervening enactment of a variety of possible exogenous incentives (e.g. national climate policy, financial support from external organizations, and diffusion of knowledge and norms). Thus, the opportunity structure is not fixed but is itself dynamic, and potentially also indeterminate and contested.

Overall patterns of empowerment/disempowerment may vary for different types of urban climate governance initiatives. Policy innovation may be realized through either active or passive empowerment. Yet this may be at risk of active disempowerment through blocking by certain powerful actors. Experimentation may be realized through passive empowerment. Yet this may be at risk of passive disempowerment if it cannot become embedded or scaled up to change the broader governance system, or active disempowerment if it is perceived to threaten powerful interests. Urban laboratories may be realized through passive empowerment. Yet this may be at most risk of passive disempowerment if they struggle to demonstrate success to involved actors over time, or confront an inability to change broader structural forces (e.g. following Evans & Karvonen, 2014). In the case of Amsterdam, evidence for both active and passive empowerment was identified, as well as a risk of active disempowerment in coming years linked to resources.

An important area requiring further work is to understand the nuances of shifts in power (e.g. specific ways in which this occurs, both formally and informally, and its consequences). This chapter starts to shed some light by clarifying differences between active and passive forms of empowerment or disempowerment. For example, there are longstanding debates in multiple fields about devolution (e.g. in environmental governance, as well as community development and social policy), and its role in producing empowerment. This chapter suggests that it is important to distinguish whether changes in decision-making are operational (e.g. for specific issues and moments in time only, such as participation in multi-actor forums) or constitutional (e.g. more permanent changes in decision-making rules). Both can potentially produce empowerment (passive or active, respectively), but with different underlying implications about shifts in power. Hence through this lens, devolution would be a sufficient but not a necessary condition for empowerment.

Strategic Use of Mechanisms

A pertinent question is, to what extent can knowledge of these mechanisms be used to inform purposeful strategies to help better realize urban climate governance initiatives? This issue needs to be further examined. Mechanisms identified may be useful in thinking comprehensively about how urban climate governance initiatives may become empowered or disempowered, and which mechanisms may be most relevant to target in a particular case. The approach also brings attention to how agency plays out within a specific opportunity structure, and the implications for shifts in power.

Realizing urban climate initiatives may really require concerted agency to take action and persuade other actors rather than expecting public demands to drive it. For example, in a recent review from the USA, Hughes (2015) found 'little evidence that urban climate change adaptation planning is happening in response to bottom-up demands; instead local governments are often developing mechanisms for engaging the public and generating interest in and support for adaptation planning'. Based on comparative case analysis in the Global South, Carmin et al. (2012) implied that empowerment requires a certain agility on the part of involved actors to adapt their initiatives and ways they are pursued within dynamic circumstances, to find opportunities for empowerment that may be linked to both endogenous and exogenous factors.

This also raises questions about empowerment in the context of broader structural factors and socio-economic development challenges that may work against it. For example, Evans and Karvonen (2014) reflected on how some urban initiatives such as urban laboratories may replicate existing patterns of social power because privileged actors benefit from the production and circulation of new knowledge while the exclusion of less privileged actors is reinforced. In their study in Bangalore, Bulkeley and Castán Broto (2014) hinted at how a lack of attention to inequity in a context marked by it could jeopardize the sustainability of the private urban initiative studied. More broadly, Hardoy and Romero-Lankao (2011) argue that the pervasive background condition of socio-economic disparity and chronic vulnerability is fundamentally disempowering to urban climate initiatives in a Latin American context.

This raises questions about how cities can seek to transform broader structural conditions in pursuing urban climate governance initiatives. In this light, an empowerment/disempowerment lens might help as a diagnostic in identifying key places where action is needed – in other words, as an entry point for targeting efforts towards transforming broader systemic structures that need to be addressed. A recent example from a high-capacity context is the case of New York City expressing its intention in January 2018 to significantly divest from pension funds investing in fossil fuel industries, and simultaneously pursue legal cases against large fossil fuel companies for climate impacts incurred in the city. This can be seen as a bold attempt at transforming broader structural conditions on urban climate action. Most cities across the world may not currently have the ability or willingness to take similar action, but this perhaps demonstrates what such purposeful attempts can look like.

Contributions to Urban Governance Scholarship

This chapter contributes to analysing the dynamics of agency and opportunity structure within urban governance systems, and the mechanisms by which specific initiatives aiming to bring about change in systems may be empowered or disempowered. The chapter contributes a critical but pragmatic analytical approach centred on urban climate initiatives. The approach is particularly oriented towards informing strategic action, in other words, ex ante efforts to design and pursue urban climate initiatives. This has potential to be further developed into a diagnostic approach in future work. However, it could also be used for ex post evaluation.

More broadly, the chapter relates to studies of power in environmental governance (e.g. Partzsch, 2017) by considering ways in which power may shift as a result of collective efforts of actors to realize new initiatives. It also relates to longstanding debates about the role of the state in environmental governance (e.g. Duit, Feindt, & Meadowcroft, 2016) which are re-emerging in recent years particularly through the frame of polycentric climate governance (Jordan et al., 2015). Urban climate governance has developed a rich body of thinking about the interplay between state and non-state actors, and provides an ideal conceptual arena for understanding change in polycentric systems as a result. Mechanisms of empowerment/disempowerment may be an important aspect of this issue. Lastly, it relates to the nascent topic of understanding transformations in governance systems (Biermann et al., 2012; Bulkeley et al., 2016; Patterson et al., 2017), and the specific mechanisms by which this may be purposefully pursued in urban settings in practice.

Acknowledgements

James Patterson gratefully acknowledges funding received from the European Union's Horizon 2020 research and innovation programme under the Marie Sklodowska-Curie grant agreement No. 659065 which supported part of this work.

References

Alsop, Ruth, Bertelsen, Mette, & Holland, Jeremy. (2005). Empowerment in practice: From analysis to implementation. *The World Bank*. https://doi.org/10.1596/978–0-8213–6450-5

Anguelovski, Isabelle, & Carmin, JoAnn. (2011). Something borrowed, everything new: Innovation and institutionalization in urban climate governance. *Current Opinion in Environmental Sustainability*, 3(3): 169–175. https://doi.org/10.1016/j.cosust.2010.12.017

Aylett, A. (2013). The socio-institutional dynamics of urban climate governance: A comparative analysis of innovation and change in Durban (KZN, South Africa) and Portland (OR, USA). *Urban Studies*, 50(7): 1386–1402. https://doi.org/10.1177/0042098013480968

Bai, Xuemei, McAllister, Ryan R. J., Beaty, R. Matthew, & Taylor, Bruce. (2010). Urban policy and governance in a global environment: Complex systems, scale mismatches and public participation. *Current Opinion in Environmental Sustainability*, 2(3): 129–135. https://doi.org/10.1016/j.cosust.2010.05.008

Bai, Xuemei, Surveyer, Alyson, Elmqvist, Thomas, et al. (2016). Defining and advancing a systems approach for sustainable cities. *Current Opinion in Environmental Sustainability*, 23(December): 69–78. https://doi.org/10.1016/j.cosust.2016.11.010

Beach, Derek, & Pedersen, Rasmus. (2016). *Causal Case Study Methods: Foundations and Guidelines for Comparing, Matching, and Tracing*. Ann Arbor, MI: University of Michigan Press. https://doi.org/10.3998/mpub.6576809

Betsill, Michele, & Bulkeley, Harriet. (2007). Looking back and thinking ahead: A decade of cities and climate change research. *Local Environment*, 12(5): 447–456. https://doi.org/10.1080/13549830701659683

Biermann, Frank, Abbott, Kenneth, Andresen, Steinar, et al. (2012). Transforming governance and institutions for global sustainability: Key insights from the Earth System Governance Project. *Current Opinion in Environmental Sustainability*, 4(1): 51–60. https://doi.org/10.1016/j.cosust.2012.01.014

Bulkeley, Harriet, Andonova, L. B., Betsill, Michele M., et al. (2014). *Transnational Climate Change Governance*. New York: Cambridge University Press.

Bulkeley, Harriet, & Betsill, Michele. (2005). Rethinking sustainable cities: Multilevel governance and the 'urban' politics of climate change. *Environmental Politics*, 14(1): 42–63. https://doi.org/10.1080/0964401042000310178

Bulkeley, Harriet, & Betsill, Michele. (2013). Revisiting the urban politics of climate change. *Environmental Politics*, 22(1): 136–154. https://doi.org/10.1080/09644016.2013.755797

Bulkeley, Harriet, & Castán Broto, Vanesa. (2013). Government by experiment? Global cities and the governing of climate change. *Transactions of the Institute of British Geographers*, 38(3): 361–375.

Bulkeley, Harriet, & Castán Broto, Vanesa. (2014). Urban experiments and climate change: Securing zero carbon development in Bangalore. *Contemporary Social Science*, 9(4): 393–414. https://doi.org/10.1080/21582041.2012.692483

Bulkeley, Harriet, Coenen, Lars, Frantzeskaki, Niki, et al. (2016). Urban living labs: Governing urban sustainability transitions. *Current Opinion in Environmental Sustainability*, 22(October): 13–17. https://doi.org/10.1016/j.cosust.2017.02.003

Burch, Sarah. (2010). Transforming barriers into enablers of action on climate change: Insights from three municipal case studies in British Columbia, Canada. *Global Environmental Change*, 20(2): 287–297. https://doi.org/10.1016/j.gloenvcha.2009.11.009

Carmin, JoAnn, Anguelovski, Isabelle, & Roberts, Debra. (2012). Urban climate adaptation in the Global South: Planning in an emerging policy domain. *Journal of Planning Education and Research*, 32(1): 18–32. https://doi.org/10.1177/0739456X11430951

Castán Broto, Vanesa, Oballa, Bridget, & Junior, Paulo. (2013). Governing climate change for a just city: Challenges and lessons from Maputo, Mozambique. *Local Environment*, 18(6): 678–704. https://doi.org/10.1080/13549839.2013.801573

Circle Economy, Copper8, & City of Amsterdam. (2018). Amsterdam Circulair. Evaluatie en Handelingsperspectieven.

City of Amsterdam. (2016). Amsterdamse City Deal Naar Een Stad Zonder Aardgas (Amsterdam City Deal Towards a City Free of Natural Gas). City of Amsterdam.

Doren, van D., Giezen, M., Driessen, P. P. J., & Runhaar, H. A. C. (2016). Scaling-up energy conservation initiatives: Barriers and local strategies. *Sustainable Cities and Society*, 26(October): 227–239. https://doi.org/10.1016/j.scs.2016.06.009

Doren, van Didi, Driessen, Peter P. J., Runhaar, Hens, & Giezen, Mendel. (2018). Scaling-up low-carbon urban initiatives: Towards a better understanding. *Urban Studies*, 55 (1): 175–194. https://doi.org/10.1177/0042098016640456

Duit, Andreas, Feindt, Peter H., & Meadowcroft, James. (2016). Greening Leviathan: The rise of the environmental state? *Environmental Politics*, 25(1): 1–23. https://doi.org/10.1080/09644016.2015.1085218

Evans, James, & Karvonen, Andrew. (2014). 'Give me a laboratory and I will lower your carbon footprint!' – Urban laboratories and the governance of low-carbon futures: Governance of low carbon futures in Manchester. *International Journal of Urban and Regional Research*, 38(2): 413–430. https://doi.org/10.1111/1468-2427.12077

Gordon, David J., & Johnson, Craig A. (2017). The orchestration of global urban climate governance: Conducting power in the post-Paris climate regime. *Environmental Politics*, 26(4): 694–714. https://doi.org/10.1080/09644016.2017.1320829

Hardoy, Jorgelina, & Lankao, Patricia Romero. (2011). Latin American cities and climate change: Challenges and options to mitigation and adaptation responses. *Current Opinion in Environmental Sustainability*, 3(3): 158–163. https://doi.org/10.1016/j.cosust.2011.01.004

Hisschemöller, M. (2016). The energetic city: Between dreams and deeds. In V. Mamadouh & A. van Wageningen (eds.), *Urban Europe: Fifty Tales of the City*. Amsterdam: Amsterdam University Press.

Hoek, J. P., Struker, van der A., & de Danschutter, J. E. M. (2017). Amsterdam as a sustainable European metropolis: Integration of water, energy and material flows. *Urban Water Journal*, 14(1): 61–68. https://doi.org/10.1080/1573062X.2015.1076858

Hughes, Sara. (2015). A meta-analysis of urban climate change adaptation planning in the U.S. *Urban Climate*, 14(December): 17–29. https://doi.org/10.1016/j.uclim.2015.06.003

Hughes, Sara. (2017). The politics of urban climate change policy: Toward a research agenda. *Urban Affairs Review*, 53(2): 362–380. https://doi.org/10.1177/1078087416649756

Jordan, Andrew J., Huitema Dave, Hildén, Mikael, et al. (2015). Emergence of polycentric climate governance and its future prospects. *Nature Climate Change*, 5(11): 977–982. https://doi.org/10.1038/nclimate2725

Marselis, Ilonka, & Hisschemöller, Matthijs. (2018). 'Het moet niet te avontuurlijk worden'. Een onderzoek naar de institutionele barrières voor een wijkgebonden warmtevoorziening in Amsterdam. Research report. DRIFT for transition, Erasmus University, the Netherlands.

Partzsch, Lena. (2017). 'Power with' and 'power to' in environmental politics and the transition to sustainability. *Environmental Politics*, 26(2): 193–211. https://doi.org/10.1080/09644016.2016.1256961

Patterson, James, Schulz, Karsten, Vervoort, Joost, et al. (2017). Exploring the governance and politics of transformations towards sustainability. *Environmental Innovation and Societal Transitions*. https://doi.org/10.1016/j.eist.2016.09.001

Roemers, G., & Faes, K. (2017). *Roadmap Circulaire Gronduitgifte: Een Introductie in Circulaire Bouwprojecten (Roadmap for the Circular Allotment of Land)*. Amsterdam: City of Amsterdam, SGS Research & Metabolic.

Savini, Federico, Boterman, Willem R., van Gent, Wouter, P. C., & Majoor, Stan. (2016). Amsterdam in the 21st century: Geography, housing, spatial development and politics. *Cities*, 52(March): 103–113. https://doi.org/10.1016/j.cities.2015.11.017

TNO, Circle Economy, & FABRIC (2015). Amsterdam Circulair: Visie & Routekaart voor stad en regio.

Voytenko, Yuliya, McCormick, Kes, Evans, James, & Schliwa, Gabriele. (2016). Urban living labs for sustainability and low carbon cities in Europe: Towards a research agenda. *Journal of Cleaner Production*, 123(June): 45–54. https://doi.org/10.1016/j.jclepro.2015.08.053

4

Transnational Municipal Networks and Cities in Climate Governance

Experiments in Brazil

FABIANA BARBI AND LAURA VALENTE DE MACEDO

Introduction

Transnational cooperation networks of local governments addressing climate change have fostered the development of an explicitly urban approach to climate governance (Schroeder & Bulkeley, 2009), beyond sectoral framing. They have provided local governments with inspiration, information, experience, capacities, concrete projects, access to funding, examples of good practices, and informal structures of recognition and rewards, which have led to significant responses to the climate change challenge by local governments worldwide. Their engagement in these transnational municipal networks (TMNs) has thus been a driver of climate action beyond co-beneficial outcomes at the local level, with great potential for the development of effective policies and actions that will contribute to address a problem with global consequences (Acuto & Rayner, 2016; Bai et al., 2010; Bulkeley, 2010; Jordan et al., 2015; Lee, 2015).

Cities engage in TMNs in search for benefits such as peer-to-peer learning, technical knowledge exchange, access to funding, and international acknowledgement and global competitiveness. Collectively, they aim to have a voice in an increasingly globalized environment. Their actions to improve health, safety, and environmental protection are enhanced and find support in partnerships forged by TMNs with global actors such as multilateral banks, agencies, and governments (Acuto & Rayner, 2016; Betsill & Bulkeley, 2004; Hickmann, 2017; Kern & Bulkeley, 2009; Toly, 2008).

Since the early 2000s, authors (Bulkeley & Betsill, 2013; Bulkeley et al., 2014; Gordon & Acuto, 2015) have recognized a 'second wave' of municipal action on climate change, characterized by a broader range of transnational networks and a growing interest in adaptation and mitigation, as well as a more political approach to urban climate governance. Historically, the main emphasis of these city networks was the local and regional development of infrastructure and energy, a concern of developed countries, where most of these municipal climate networks are active (Bulkeley et al., 2014). Individual city-based action is now standard in much of the world, though perhaps more in the Global North than in the Global South (Bernstein & Hoffmann, 2018). Some of these networks are not representative or even active in South America, such as Climate Alliance and Energy Cities. Besides, individual city-based climate action is not standard in South American cities, and literature on TMNs is strongly based on analysis of Global North cities and networks (see Bansard et al., 2017).

TMNs have also been responsible for undertaking climate experiments with local governments aiming mainly at greenhouse gas (GHG) emissions reduction. Experimentation, understood as initiatives outside the state-based multilateral treaty (Hoffmann, 2011), involving innovation and learning (Bulkeley & Castán Broto, 2013), represents a critical aspect of urban responses to climate change. Thus, there is the need to understand these interventions, led by local governments, TMNs, and other agents in pursue of low-carbon cities, building on technology and new institutional arrangements. Scholarly debate has begun to hone in on such experiments, mostly in the Global North (Bulkeley et al., 2015; Caprotti & Cowley, 2017; Castán Broto & Bulkeley, 2013; Raven et al., 2017).

Climate governance experiments have grown along with the growth of TMNs in number, size, and visibility. For instance, by 2018 Climate Alliance had 1,719 city members in 26 countries[1]; ICLEI – Local Governments for Sustainability (formerly named International Council for Local Environmental Initiatives)1,500 members in 84 countries[2]; the United Cities and Local Governments (UCLG) 1,000 members in 140 countries[3]; C40 Climate Leadership Group 96 members in 56 countries[4], among others. ICLEI was the first and is the largest transnational network of local governments engaged in climate action through its international campaign Cities for Climate Protection (CCP) launched in 1993 (Harvey, 1993; ICLEI, 1997, 2017; Lambright et al., 1996).

ICLEI established the Latin America and the Caribbean Secretariat (LACS) in Brazil in 2001, and since then it has increased its membership in the country, from 12 to more than 50 municipalities in 17 years, mainly through the CCP, which was inaugurated in South America in 2002. Many climate initiatives regarded as innovative in Brazil were developed in the context of the CCP or within one of its spin-off projects and programmes.

In this chapter, we examine the CCP in Brazil to determine how TMNs have enabled local governments to undertake more experimental forms of climate governance and intervention.

[1] www.climatealliance.org/nc/municipalities/the-network.html (accessed 7 May 2018).
[2] www.iclei.org/iclei-members/iclei-members.html (accessed 7 May 2018).
[3] www.uclg.org (accessed 7 May 2018). [4] www.c40.org/cities (accessed 7 May 2018).

This chapter is divided into two sections. First, we discuss the role of TMNs in fostering experimentation in climate governance. Next, we assess ICLEI's CCP implementation in Brazilian cities by mapping climate change experiments throughout the campaign's three phases. By establishing the connection between TMNs as collective agents and urban climate experimentation in Brazilian cities, this chapter aims to contribute to the literature on urban climate governance taking place in the Global South.

The Role of TMNs and Cities' Experimentation in Climate Governance

There is enough evidence in the literature (Bellinson, 2018; Betsill & Bulkeley, 2004, 2007; Bouteligier, 2015; Fünfgeld, 2015; Hickmann, 2017; Rashidi & Patt, 2018; Toly, 2008) that TMNs play an important role in climate governance worldwide, mainly by motivating cities to define and develop climate policy initiatives and by helping them implement these strategies. Most of these studies have focused on the relevance of city networks to the adoption of climate policies, mainly by analysing the reasons cities join these networks (Betsill & Bulkeley, 2004; Buis, 2009; Heinrichs et al., 2013; Kern & Bulkeley, 2009; Pitt, 2010). Other studies address the impact that networks have on the policies adopted (Hickmann et al., 2017; Krause, 2012; Lee & Koski, 2014; Rashidi & Patt, 2018).

Few studies have addressed the role of TMNs in fostering experimentation in climate governance in the cities (Bellinson, 2018; Castán Broto & Bulkeley, 2013; Fünfgeld, 2015; Gordon, 2013). Climate governance experimentation has been explored over the last 10 to 15 years, with subnational actors and initiatives at the vanguard (e.g. Andonova et al., 2017; Bulkeley et al., 2014; Hoffmann, 2011).

Subnational climate governance experimentation can be referred to as experiments initiated by both subnational governments and non-state actors acting at the subnational level, in defiance of the state-led multilateral climate treaty (Bernstein & Hoffmann, 2018; Hoffmann, 2011). The first climate governance experiments recognized in the literature were municipal climate action and TMNs (Betsill & Bulkeley, 2004).

As subnational experimental climate governance engaged diverse agents and new political-institutional arrangements, authors focused on understanding its origin, functioning, organization, and whether it could provide effective climate responses (Bernstein & Hoffmann, 2018; Bulkeley et al., 2014; Hoffmann, 2011; Jordan et al., 2015; Meckling et al., 2015). Less attention has been directed towards understanding the interventions or experiments through which climate change responses are being organized and pursued in cities (Bulkeley et al., 2014).

There is no consensus definition of climate governance experimentation or experiments; however, scholars agree that the concept involves something new being tried out, an intervention differing from the status quo, beyond the UN treaty. Therefore, a constitutive element of experimentation is novel action. Some authors (e.g. Voytenko et al., 2016) analyse technological innovations, pilot initiatives, policy experiments, or urban living labs. Scholars from the Earth System Governance project have focused on novel sources of rulemaking and authoritative actors (Bernstein & Hoffmann, 2018; Castán Broto & Bulkeley, 2013; Hoffmann, 2011). In these cases, climate governance experimentation implies diverse forms of technical and social innovation, and trial and error with new forms of governance only loosely or not connected to the traditional forms of governance.

In climate-related experimentation, the objective is to induce change in current unsustainable practices in order to mitigate climate change (Matschoss & Repo, 2018). The consensus among these studies is that climate governance experiments can provide opportunities to identify novel ways to address the climate change challenge and provide improved possibilities for transition to low-carbon societies.

It has been argued that experimentation is a critical aspect of urban responses to climate change where local governments and private and civil society actors seek to experience and learn what might mean responding to the climate change challenge through a variety of interventions, projects, initiatives and arrangements (Bulkeley & Castán Broto, 2013; Bulkeley et al., 2015). These authors argue that climate change experiments are integral to the governing of climate change in cities. Their studies have shown that such experiments are not only cases of good practice, but also have importance and potential for climate change governance.

In this chapter we consider urban climate governance experiment an initiative that addresses climate change with the intention to set up rules that shape climate change responses at the city level. The novel element is in the rulemaking process, as well as the issue itself being addressed at the municipal level.

The understanding of the relation between TMNs and climate governance experimentation is largely derived from case study work, predominantly of cities in the Global North and on mitigation responses (Bouteligier, 2015; Castán Broto & Bulkeley, 2013; Edwards & Bulkeley, 2018; Hickmann, 2017; Hoffmann, 2011; Matschoss & Repo, 2018). Few studies investigated cases in the Global South, and even fewer in Latin American and Brazilian cities (Bulkeley et al., 2015; Castán Broto & Bulkeley, 2013; Hickmann et al., 2017).

Some studies surveyed a sample of several cities at the same time: Castán Broto and Bulkeley (2013) catalogued 627 municipal climate experiments in 100 cities, uncovering the heterogeneous mix of agents, settings, governance arrangements, and technologies involved in the governance of climate change in cities in different parts of the world; Matschoss and Repo (2018) examined 141 climate action experiments from 28 European Union countries, comparing governance innovations against other kinds of innovations (organizational, product, service, social and system).

Even though these studies bring important contribution to understand the relation between TMNs and climate governance experimentation, the literature lacks elements to understand the role played by TMNs in climate governance in Brazilian cities and how they enabled climate experiments in those localities.

Cities for Climate Protection in Brazil

Since the early 1990s, Brazilian municipal governments have been increasingly engaging in international cooperation activities, beyond twinning initiatives and projects, participating in networks such as the UCLG, Metropolis, WHO′s Healthy Cities, and establishing their own international relations departments.

Scholarly debate has focused on the political, legal and institutional aspects of subnational government internationalization in Brazil (Aprigio, 2016; Macedo et al., 2016; Salomón, 2011; Setzer, 2009, 2013; Vigevani, 2006). The growing participation of Brazilian cities in TMNs has opened new avenues for research on the empowerment of local governments vis-à-vis global climate governance. However, the role and impact of such networks in the Brazilian context is still poorly understood. Furthermore, the effectiveness of current institutional architecture for climate governance in Brazil is yet to be investigated in depth.

Brazilian political system is strongly based on normatization founded on Roman law; establishing a legal framework is a key step for designing and implementing public policies at a subnational level – thus the importance of such initiatives regarding specific climate policies, beyond pilot projects or geographically bound initiatives limited in scope and time. In the case of local climate policies, participating in TMNs was the driver for cities to undertake climate governance experimentation and aim for continuity through rulemaking.

We argue that climate change interventions in Brazilian cities are driven by TMNs and they happen at the designing phase, whereas implementation is compromised without TMNs' support. Therefore, climate experiments may not necessarily mean GHG emission reduction or effective adaptation to climate change impacts. Climate governance experiments fostered by TMNs are, given coordination and authority constraints, subjected to problems related to implementation (Bulkeley, 2010; Gordon, 2013). This gap limits the governance potential of TMNs, as they have minimal tools regarding monitoring and oversight (they are largely reliant on self-reporting by city members) and enforcement. Also, the

Figure 4.1 CCP timeline in Brazil.

implementation gap may compromise the potential of climate governance experiments to catalyse decarbonization trajectories.

Among the city networks active in Brazil, ICLEI's CCP was the first to address climate change and the one that engaged most Brazilian cities. The purpose of the Campaign was to facilitate urban emissions reduction through a five-milestone process: (1) measurement (emissions inventory), (2) commitment (target establishment), (3) planning (develop climate action plan), (4) implementing and (5) monitoring/evaluating progress.[5]

Studies about CCP cities in Asia, Australia, the United States, and Europe analyse examples of ongoing initiatives and projects to conceptualize multilevel climate governance (Bansard et al., 2017; Castán Broto & Bulkeley, 2013; Hickmann, 2017; Krause et al., 2016; Schroeder & Bulkeley, 2009; Yienger et al., 2002) but few studies focused on CCP's actions in South American cities (Barbi & Ferreira, 2017; Macedo, 2017).

In Latin America, the CCP campaign began in Mexico, in 1998, and was extended to South America in 2002, funded by the Canadian International Development Agency (CIDA). The agenda focused on mitigation and successfully recruited cities in Argentina, Brazil, and Chile.[6] In Brazil, CCP's actions can be divided into three distinct phases, determined both by the international CCP's agenda and funding (Figure 4.1).

TMNs generally focus on some of the following 21 issue areas: renewable energy, energy efficiency, transport, clean fossil fuels, carbon markets, carbon finance, demand reduction, waste, water use, carbon sequestration, forests and biodiversity, sea levels, water shortage, agriculture and food, disaster risk reduction, extreme weather, flood risk, health, insurance and risk, resilience, and heat stress (Bulkeley et al., 2014: 86). In Brazil, CCP projects targeted transport, sustainable procurement and construction, waste management and urban sustainability planning throughout their implementation. Issue areas related to adaptation emerged mostly in CCP's third phase.

[5] http://archive.iclei.org/index.php?id=10829 (accessed 23 April 2017).
[6] www.iclei.org/index.php?id=1768 (accessed 30 July 2010).

First Phase of CCP (2001–2005): Breaking Ground for Climate Experiments

In its first phase, the CCP was presented as a campaign, and along with Agenda 21 provided the framework for ICLEI's actions with Brazilian municipalities. Rio de Janeiro was the first host city for the Latin American and the Caribbean Secretariat (ICLEI-LACS), and CCP in South America was then a pilot project, concentrated in Brazil. As such, the project helped anchor ICLEI in the region, while introducing climate change in the municipal agenda. At the time, cities in Brazil were not engaged or even interested in global climate change issues, addressed mostly by federal government and the epistemic community.

CCP placed local actors in the climate governance debate when they still did not have GHG emission reduction commitments (Schroeder & Bulkeley, 2009; Yienger et al., 2002). Therefore, this phase was important to set the stage for climate experiments in the period that followed. During this phase, CCP brought a strong focus on outreach and awareness building in the seven Brazilian cities that adhered to the campaign: Betim, Goiânia, Palmas, Porto Alegre, Rio de Janeiro, São Paulo and Volta Redonda.

ICLEI assisted the cities in developing GHG emissions inventories, establishing reduction targets, and identifying priority actions to reduce their emissions within the five-milestone methodology. Solid waste management and urban transport were identified as priority issues in this first phase; to a lesser extent, renewable energy was also a concern. Vulnerability and adaptation were not addressed by the cities at the time. Table 4.1 synthesizes CCP's actions in its first phase.

Second Phase of CCP (2005/6–2011): Two Sets of Climate Experiments

During the second phase, after the pilot project was concluded, the CCP became an umbrella programme, with sectoral projects funded by different sources. Activities now focused on milestone 4: other issues related to GHG emissions reduction were added to the agenda, such as public procurement, sustainable construction, renewable energy and urban resilience. Besides CCP members, these projects involved cities such as Belo Horizonte, Campinas, Manaus and Santo Andre, as well as states and federal government departments.

During this phase, state governments with political and economic importance, such as São Paulo, Minas Gerais, Rio de Janeiro, Bahia, Mato Grosso and Pernambuco adhered to ICLEI initiatives. In Brazil, metropolitan areas fall within state jurisdiction, therefore state governments were considered as local governments for the purpose of engaging them in specific projects. The strategy was

Table 4.1 *CCP timeline and milestones in Brazil (First Phase 2001–2005)*

First Phase – Pilot project – CCP Campaign in South America				
2001	2002	2003	2004	2005
CIDA funding for CCP South America, through ICLEI USA	Selection process establishes nine participating cities. June: CCP launched in workshop in Volta Redonda, RJ.	Palmas and São Paulo join; Niteroi leaves.	All cities reported initiatives to reduce GHG emissions (milestone 4).	February: Kyoto Protocol enters into force. CCP pilot project with CIDA funding closed in March.
MoU signed between ICLEI and Rio to host LACS.	Rio de Janeiro, Volta Redonda, Niterói, Betim, Porto Alegre, Goiânia, Buenos Aires, Avellaneda (AR) and Tomé (CL).	Focus on Solid Waste Management. Workshop in Rio de Janeiro.	Focus on urban transport. Workshop in São Paulo.	Spin-offs: Renewable Energies project and Sustainable Public Procurement project – FCO-UK funding. States of SP and MG participate and join ICLEI network.
Invitation letters sent to 40 SA cities, 14 responses with expression of interest.		All cities complete milestones 1, 2, and 3.	COP10 in Buenos Aires. International seminar on CDM opportunities for cities.	CCP mentioned in Brazil´s first national communication to the UNFCCC, as long-term action for climate mitigation.

Source: Authors, adapted from Macedo (2017).

instrumental to promote vertical integration and strengthen local government vis-à-vis climate governance in Brazil. At the time, engaging Brazilian state governments was a pioneer initiative in ICLEI; later, other regional offices recruited provinces and states. Presently, many ICLEI led international projects involve regions and provinces, such as the 100%RE Cities & Regions Network[7] and the SSP Regions[8] (ICLEI, 2017).

Two sets of climate experiments can be highlighted during the second phase. The first one addresses sustainable consumption as a means to contribute to GHG emission reduction. This climate experiment was developed within two projects: Sustainable Public Procurement (SPP) and Sustainable Building Policies (PoliCS).

Begun in 2006, the SPP campaign became an important area of expertise within ICLEI-LACS, modelled after the PROCURA+ campaign in Europe. ICLEI-LACS had been negotiating the initiative since 2003, and with funding from the British government developed customized reference materials and manuals adapting ICLEI's SPP methodology to the Brazilian legislation and context. Capacity building for audiences in all levels of government was a key feature. Several initiatives promoting sustainable public procurement were implemented in cities, states, Brazilian congress, ministries, the federal Attorney General´s Office, among other departments, with ICLEI guidance.

PoliCS tackled GHG emissions in the building sector. Participating cities developed and implemented regulation on timber, energy efficiency and renewable energy. The project outputs included reference materials and capacity-building activities.

Both SPP and PoliCS addressed deforestation, a priority issue in Brazil as timber coming from illegal logging in the Amazon contributes to land use change. This sector was the main source of Brazilian emissions during most of the period between 1970 and 2016 (Brasil, 2017; SEEG, 2017). An estimate based on official data from 2004 accounted that 59 per cent of the timber from the Amazon region was harvested illegally; about 42 per cent was consumed in the south and southeast of Brazil, mostly in the state of São Paulo, for construction (Greenpeace, 2005). During this period, the city of São Paulo issued an ordinance on public procurement of timber to monitor the supply chain, and verify its origin, in order to guarantee use of legally harvested wood. The Environment Secretary (SVMA) led the initiative, which had been instituted previously. With ICLEI´s support, it gained traction and integrated other departments in the administration. A local experiment thus grew and scaled up, inspiring other governments' purchasing decisions, including the states of Bahia, Rio de Janeiro, and Pernambuco.

The second set of climate experiments regards climate change policies. Subnational governments focused on strengthening the legal and institutional framework for climate protection. Three Brazilian cities and three states developed and approved their climate policies within this phase of CCP: the cities of Belo

[7] https://iclei.org/en/100RE.html (accessed 7 January 2019).
[8] www.sppregions.eu/ and www.procuraplus.org (accessed 7 January 2019).

Horizonte, Rio de Janeiro and São Paulo and the states of Bahia, Mato Grosso, and Pernambuco. Other cities and the states of Rio de Janeiro and São Paulo followed suit. These experiments were fostered by two CCP-related projects: PAVS ('Green and Healthy Environment') and PeClima ('Developing State Policies and Action to Combat Climate Change').

PAVS was an initiative of the city of São Paulo to establish the climate–health nexus. A main feature of the project was to integrate municipal departments by strengthening institutional capacity and improving environmental management through the climate policy. ICLEI played a key role helping the city to formulate the Bill on the Municipal Climate Change Policy, submitted by the Executive to the Legislative and passed in June 2009 by City Council, in a historic unanimous vote. Besides technical staff and leadership from several municipal departments, drafting the policy involved experts in academia; practitioners from relevant business sectors, such as construction, transport, energy and waste management; non-government organization (NGO) representatives; and officials from other levels of government. Not only was the theme novel, but the policy design process innovated by integrating the municipal administration and civil society, represented by the NGO, business and epistemic communities (Biderman, 2011).

São Paulo's municipal climate policy was a pioneer in establishing a GHG emissions reduction goal in Brazil. It set the target to reduce its emissions by 30%, until 2012, based on 2005. This influenced the State of São Paulo's and the national policies to be approved and set reduction targets, later in 2009. Some studies (Biderman, 2011; Macedo, 2017; Setzer, 2009; Vargas, 2009) indicate that the participation of São Paulo in TMNs and the commitments the city assumed within these networks were crucial for the city to adopt a climate policy with concrete targets.

However, by 2012, net GHG emissions in São Paulo had increased slightly, mainly due to the growth in the city's car fleet (Geoklock & Ekos Brazil, 2012), driven by federal fiscal policies that favoured the auto and oil industries. Actors in government and NGOs, politicians, and experts were very critical – and very vocal – about the city's ambition in establishing a short-term goal.[9] What was conveyed by adversaries through media as a mistake, provided justification for the ensuing administration to ignore the Law and paralyse many of its initiatives (Setzer et al., 2015). The failure to meet the target however, did not imply that the policy was ineffective. Given its limited jurisdiction over energy sources, one of the biggest challenges for the municipality is reducing GHG emissions. The process itself, viewed as an experiment in climate governance involved trial

[9] Articles in the media: https://exame.abril.com.br/mundo/sao-paulo-nao-atinge-meta-de-reducao-de-emissoes/; www.ebc.com.br/noticias/brasil/2013/03/sao-paulo-nao-atinge-meta-de-reduzir-emissoes-de-gas-de-efeito-estufa-aponta; www.valor.com.br/brasil/3042928/cidade-de-sao-paulo-descumpre-meta-de-reduzir-emissoes-em-30 (accessed 7 January 2019).

Table 4.2 *CCP timeline and milestones in Brazil (Second Phase: 2005–2011)*

Second Phase: CCP Programme in Brazil					
2005/6	2007	2008	2009	2010	2011
ICLEI-LACS is hosted by Buenos Aires. City of São Paulo hosts a local office for Brazil.		Local Government Climate Roadmap launched in Bali during COP13.	ICLEI-LACS is hosted by the City of São Paulo.	Sustainable construction initiatives with the state of Rio de Janeiro.	São Paulo hosts C40 international summit
Implementation of thematic projects on renewable energies. SPP guide launched in Portuguese	Partnership with São Paulo to develop specific mitigation projects under PAVS.	Sustainable construction project – Polics – building green cities.	PEClima: Climate action with state governments of Bahia, Mato Grosso and Pernambuco.	Sustainable public procurement initiatives with the state of Bahia and federal government.	
Partnerships on SPP developed with cities.	Projects on sustainable construction: Polics with cities.		Cities and states participate in COP15, Copenhagen	ICLEI establishes partnership with IUCN to develop biodiversity project for cities	

Source: Authors, adapted from Macedo (2017).

and error and learning (Bernstein & Hoffmann, 2018; Bulkeley & Castán Broto, 2013; Castán Broto & Bulkeley, 2013; Hoffmann, 2011). Furthermore, the timeframe for analysing the policy's success (or lack thereof) is inadequate. It will possibly take more than a decade of consistently measuring emissions to evaluate the impact of São Paulo's municipal climate policy.

Regarding the state level, PeClima was aligned with CCP, supporting the development of climate state policies in Brazil. The rationale for the project was that the state level is best suited within the federative structure to link different instances of policies and actions. The participating states established legal and institutional frameworks for climate action, and their role in implementing the national climate policy was emphasized. Thus, this phase of the CCP successfully supported scaling up urban climate experimentation. CCP's main actions in this second phase are shown in Table 4.2.

Climate experiments are conceptualized as heuristic processes, involving learning from mistakes and difficulties (Bernstein & Hoffmann, 2018; Bulkeley & Castán Broto, 2013; Castán Broto & Bulkeley, 2013; Hoffmann, 2011). Within these two sets of climate governance experiments in Brazil, there was at least on lesson learned that helped shape the next CCP phase: political commitment is crucial to the projects' success (Hickmann, 2017; Macedo, 2017). That is the reason why its follow-up campaign Green Climate Cities (GCC)'s methodology starts with commitment and mobilization. By having leaders commit publicly in a signing ceremony, the project provides political visibility and accountability, and ensures political engagement, thus avoiding setbacks.

Third Phase of CCP (2011/12–Present): Moving Forward with Experiments

The third phase of the CCP in Brazil began in 2011/2012, when the international campaign underwent a major transition, becoming the Green Climate Cities (GCC) Programme. The CCP methodology was improved to integrate the low carbon perspective to urban development strategies. The GCC offers access to tools, instruments, best practices and management support through a methodology encompassing nine steps grouped under three action areas: (i) Analyze, (ii) Act, and (iii) Accelerate. The steps are: (1) Commit and mobilize, (2) Research and access, (3) Set baseline, (4) Develop strategy, (5) Detail and finance projects, (6) Implement and monitor, (7) Integrate and collaborate, (8) Review and upscale and (9) Advocate and inspire.

In Brazil, the project that inaugurated this phase was Urban LEDS (Urban Low Emissions Development Strategy), spanning two periods: 2011–2015 and 2017–2020. Its main objective was to set a roadmap for the cities' transition towards a low-emission, green, and inclusive urban economy, by integrating it to the cities' development plans and processes. Eight Brazilian cities joined the project: Betim, Belo Horizonte, Curitiba, Fortaleza, Porto Alegre, Recife, Rio de Janeiro and Sorocaba. In the first period (2012–2015), the cities reached the fourth step in the methodology – they committed, researched and assessed, set a baseline and developed a strategy. In most cases, action focused on energy issues. Belo Horizonte, Rio de Janeiro and São Paulo had previously set targets in their climate policies (Barbi & Ferreira, 2017; Macedo, 2017). The other cities decided to set their targets based on studies conducted within Urban LEDS.

One of the major challenges facing TMNs is to find ways of scaling climate experiments to achieve meaningful effect (Hoffmann, 2011). That means making

Table 4.3 *ICLEI's methodologies for climate action, 1993–2018*

CCP – Cities for Climate Protection	GCC – Green Climate Cities	
Milestones	Phase	Steps
1. Undertake GHG baseline inventory **2.** Define voluntary reduction goal	**I** **Analyse**	**1.** Commit and mobilize **2.** Research and access **3.** Set baseline
3. Develop climate action plan **4** Implement action plan	**II** **Act**	**4.** Develop strategy **5.** Detail and finance projects **6.** Implement and monitor
5. Monitor implementation progress and restart the process.	**III** **Accelerate**	**7.** Integrate and collaborate **8.** Review and upscale **9.** Advocate and inspire

Source: ICLEI website, adapted by the authors.

sure climate experiments related to mitigation lead to actual GHG emissions reduction (Bernstein & Hoffmann, 2018). Many climate experiments did not evolve in Brazilian cities because the projects in which they were inserted ended, as well as funding. Therefore, in order to secure continuity for the projects' activities without ICLEI's support, GCC methodology included the financial aspect. In the second period of Urban LEDS, beginning in 2017, ICLEI has assisted the same cities in the implementation and funding agenda. The differences between the CCP and the GCC methodologies are shown in Table 4.3.

The GCC programme building on the CCP Campaign continues to focus on climate change mitigation. However, adaptation and urban resilience have also been gaining ground. The programme addresses challenges and opportunities of urban growth, exploring green economy and green infrastructure, and pursuing low-emission development paths of urban communities. The main idea of the programme is to integrate climate change, both mitigation and adaptation aspects in urban planning. The scope became broader, and issues such as biodiversity became part of the climate agenda.

The more comprehensive GGC methodology may foster urban climate experiments focused also on adaptation. While some studies (Bellinson, 2018; Fünfgeld, 2015) have shown the growth of climate adaptation experiments in the Global North, they are still poorly documented as such in the Global South.

Another important action led by TMNs during this phase was recruiting cities to sign the Global Covenant of Mayors for Climate and Energy (GCMCE). The GCMCE emphasizes the importance of climate change mitigation and

Table 4.4 *CCP timeline and milestones in Brazil (Third Phase: 2011/12–present)*

Third Phase: CCP develops into the Green Climate Cities Programme				
2011/12	2013	2014	2015	2016–2017
ICLEI World Congress in Belo Horizonte defines eight urban agendas in South America	SPP and innovation project with two Brazilian Ministries (British Embassy Funding)	Partnership with CB27	Global Covenant of Mayors for Climate and Energy	Urban LEDS starts second period
Urban LEDS initiated (European Commission funding)	ICLEI-LACS is divided in two centres: in Mexico, for Central America and the Caribbean; in São Paulo, for South America (ICLEI-SAMS)	Urban LEDS cities present GHG inventories (using HEAT and reported in the GPC methodology)	Urban LEDS ends first period	ICLEI SAMS supports CB27 as executive secretariat

Source: Adapted from Macedo (2017).

adaptation, as well as of increased access to clean and affordable energy. Checking its list of signatories demonstrates how climate change has gained scale in Brazil, at least in terms of awareness: until April 2017, 40 Brazilian cities had committed to climate change actions; a year later, other 17 cities had joined the agreement. ICLEI-SAMS (South America Secretariat) led the effort in coordination with other TMNs and working with other Brazilian agents, such as the National Front of Mayors (FNP), the National Association of Municipal Environmental Agencies (Anamma) and the Capital Cities Forum of Environmental Authorities (CB27). Starting in 2017, the cities must fulfil six steps in three years: GHG emissions inventory, target setting, mitigation plan, risk assessment, vulnerability assessment and adaptation plan. ICLEI offers capacity building to participating cities. Table 4.4 summarizes the third phase.

Table 4.5 synthetizes climate action led by ICLEI comparing the main characteristics of the CCP's three phases in Brazil.

Table 4.5 *Comparative table on characteristics of ICLEI's CCP three phases of in Brazil*

ICLEI CCP in Brazil	First phase (2001–2005)	Second phase (2005/6–2011)	Third phase (2011/12–2018)
Main features	CCP framed as a campaign; five-step methodology	CCP framed as an umbrella programme covering other projects	From CCP to GCC; nine-step methodology
Main focus	Mitigation. Assist cities to make their GHG emissions inventory and identify priority areas for intervention	Mitigation. Direct action and target setting through specific projects	Mitigation and adaptation, resilience. Support local communities that are on the front lines addressing the challenges and opportunities of urban growth, exploring their green economy and green infrastructure, and pursuing a low-emission development trajectory
Issue areas	Sustainable Public Procurement; Transport; Renewable energies	Sustainable public procurement; renewable energies; sustainable construction; resilient and low carbon cities; solid waste management; urban mobility.	Sustainable urban planning Urban biodiversity and conservation Energy efficiency
Projects developed	Rede Elo CPS Brasil	SPP projects Polics M2 M GeRes PAVS Peclima	Urban LEDS Global Covenant of Mayors for Climate and Energy InterAct Bio Protected urban land Building efficiency accelerator
Cities involved	Betim, Goiania, Palmas, Porto Alegre, Rio de Janeiro, São Paulo, Volta Redonda	Belo Horizonte, São Paulo, Porto Alegre, Campinas, Santo Andre, Manaus	Recife, Fortaleza, Rio de Janeiro, Belo Horizonte, Betim, Sorocaba, Porto Alegre, Curitiba, Londrina

Table 4.5 (*cont.*)

ICLEI CCP in Brazil	First phase (2001–2005)	Second phase (2005/6–2011)	Third phase (2011/12–2018)
			+40 cities of Global Covenant of Mayors for Climate and Energy
States involved	São Paulo, Minas Gerais	Bahia, Mato Grosso, Pernambuco	NA
Main outcomes	GHG Inventories; Center of Renewable Energy in Betim and Porto Alegre	GHG inventories; Climate change policies at subnational level.	GHG inventories; Climate change policies

Some TMNs governance functions can be defined as capacity building, information sharing, target setting, direct action, monitoring and certification, funding, and rule setting (Bulkeley et al., 2014). According to the data presented in this section, ICLEI's CCP main functions observed in Brazilian cities are:

1. Direct action and target setting: most projects focus on policy development
2. Support to policy development and implementation: policy recommendations and guidelines developed in accordance with the city′s reality and context
3. Peer-to-peer exchange activities for local governments: through projects with model and satellite cities, and technical visits to other cities in the ICLEI network
4. Dissemination of information to governments: portals, manuals, case studies
5. Capacity building for governments: seminars, workshops and training courses

By 2017, 11 capital cities with GHG inventories and climate policies were ICLEI members. Together they represented a population of 33,326,765 inhabitants in 2016.[10] By 2018, seven Brazilian municipalities out of 5,570 had passed legislation related to climate change. Their engagement in ICLEI's CCP played a decisive role in the adoption of their climate policies, which can be configured as climate experiments (Barbi, 2015; Barbi & Ferreira, 2017; Macedo, 2017; Martins & Ferreira, 2011; Setzer et al., 2015).

[10] Palmas, Fortaleza, Recife, Salvador, Goiânia, Belo Horizonte, Rio de Janeiro, São Paulo, Curitiba, and Porto Alegre: capitals with a GHG emissions inventory and/or climate policies that are international commitments signatories.

Final Comments

TMNs have fostered experimentation in climate governance in the cities. Subnational actors and initiatives are at the frontline of climate governance experimentation. These experiments are understood as an essential aspect of urban responses to climate change and integral to climate governance in cities.

Climate governance experiments led by cities and TMNs are potential means to catalyse decarbonization trajectories as they involve novel ways to address climate change in cities. Climate experimentation suggests social-technical innovation, also involving trial and error with new forms of governance.

In this chapter, we examined how ICLEI's CCP has enabled Brazilian local governments to create capacity to govern climate change and to undertake climate governance experiments. These experiments fostered by their paradiplomatic activities also strengthened municipalities as stakeholders in the national climate policy arena. There is no empirical evidence, however, that local governments will become agents in Brazilian climate governance in the short term, despite their recognition as players in implementing low-carbon strategies.

ICLEI has placed Brazilian cities in the climate change debate by bringing the issue to their political agenda and has driven climate governance experimentation in Brazilian cities. However, they happen more at the policy design stage, while implementation is compromised without ICLEI's support. Furthermore, the impact of subnational climate policies on urban GHG emissions cannot be determined without monitoring and verification of consistently measured and reported quantitative data. So far, efforts to do so have been sparse, as demonstrated by self-declared action in reporting platforms.[11] Therefore, climate experiments in these cities may not necessarily be leading to significant – or even any at all – GHG emissions reduction.

This scenario may change with improvements made to the CCP/GCC methodology regarding political engagement and mobilization and ensuring financial support to continue the experiments without ICLEI's presence. These improvements may help to reduce the implementation gap of such experiments in Brazilian cities.

Urban experimentation in Brazil is not new, as demonstrated by Brasília and Curitiba decades before climate change became integrated in the municipal agenda. However, further research is needed in order to understand what climate governance experiments are accomplishing and where they might be headed in Brazilian

[11] Such as Non-State Actor Zone for Climate Action (NAZCA), Disclosure Insight Action (CDP – formerly the Carbon Disclosure Project), and carbon Climate Registry (cCR).

cities. This is necessary to assess if they are possible pathways that will lead to a low-carbon transition.

The question that remains is in what ways these climate governance experiments may link with UN climate negotiations and create potential for fruitful future developments. Therefore, more detailed understanding of how and why these initiatives work or not is still necessary.

Acknowledgements

We thank Rafael d'Almeida Martins (UNDP Honduras), Bruna Cerqueira (ICLEI), and Sophia Picarelli (ICLEI) for their contributions to this chapter and the reviewers for their comments that greatly improved the manuscript.

References

Acuto, M., & Rayner, S. (2016). City networks: Breaking gridlocks or forging (new) lock-ins? *International Affairs*, 92(5): 1147–1166.

Andonova, L. B., Hale, T. N., & Roger, C. B. (2017). National policy and transnational governance of climate change: Substitutes or complements? *International Studies Quarterly*, 61(2): 253–268.

Aprigio, A. (2016). *Paradiplomacia e interdependência. As cidades como atores internacionais*. Rio de Janeiro: Gramma.

Bai, X., Roberts, B., & Chen, J. (2010). Urban sustainability experiments in Asia: Patterns and pathways. *Environmental Science and Policy*, 13(4): 312–325.

Bansard, J. S., Pattberg, P. H., & Widerberg, O. (2017). Cities to the rescue? Assessing the performance of transnational municipal networks in global climate governance. *International Environmental Agreements*, 17(2): 229–246.

Barbi, F. (2015). *Mudanças climáticas e respostas políticas nas cidades*. Campinas, Brazil: Editora da Unicamp.

Barbi, F., & Ferreira, L. C. F. (2017). Governing climate change risks: Subnational climate policies in Brazil. *Chinese Political Science Review*, 2(2): 237–252.

Bellinson, R. G. (2018). Connecting the dots: The politics of governing urban climate adaptation innovations through transnational municipal networks. In S. Hughes, E. K. Chu, &, S. G. Mason (eds.), *Climate Change in Cities: Innovations in Multi-Level Governance*. The Urban Book Series. Cham, Switzerland: Springer International Publishing.

Bernstein, S., & Hoffmann, M. (2018). The politics of decarbonization and the catalytic impact of subnational climate experiments. *Policy Sciences*. https://doi.org/10.1007/s11077-018–9314-8

Betsill, M., & Bulkeley, H. (2004). Transnational networks and global environmental governance: The cities for climate protection program. *International Studies Quarterly*, 48(2): 471–493.

Betsill, M. M., & Bulkeley, H. (2007). Looking back and thinking ahead: A decade of cities and climate change research. *Local Governments*, 12(5): 447–456.

Biderman, R. (2011). *Limites e alcances da participação pública na implementação de políticas subnacionais em mudanças climáticas e o município de São Paulo*. PhD thesis, Escola de Administração de Empresas da Fundação Getúlio Vargas, São Paulo.

Bouteligier, S. (2015). Multinational companies and urban climate governance: Market making or successful policy innovation? In J. Craig, N. Toly, & H. Schroeder (eds.), *The Urban Climate Challenge – Rethinking the Role of Cities in the Global Climate Regime*. London: Routledge, 82–99.

Brasil. (2017). *Second Biennial Update Report of Brazil to the United Nations Framework Convention on Climate Change*. Brasília, DF: Ministério de Ciência, Tecnologia, Inovações e Comunicações (MCTIC). http://sirene.mcti.gov.br/documents/1686653/2091005/BUR2-PORT-02032017_final.pdf/91e855c5-0e73-41e0-9df2-1fc0b723a51d (accessed 7 Janaury 2019).

Buis, H. (2009). The role of local government associations in increasing the effectiveness of city-to-city cooperation. *Habitat International*, 33(2):190–194.

Bulkeley, H. (2010). Cities and the governing of climate change. *Annual Review of Environment and Resources*, 35(1): 229–253.

Bulkeley, H., & Betsill, M. M. (2013). Revisiting the urban politics of climate change. *Environmental Politics*, 22(1): 136–154.

Bulkeley, H., & Castán Broto, V. (2013). Government by experiment? Global cities and the governing of climate change. *Transactions of the Institute of British Geographers*, 38 (3): 361–375.

Bulkeley, H., Castán Broto, V., & Edwards, G. (2015). *An Urban Politics of Climate Change: Experimentation and the Governing of Socio-technical Transitions*. Abingdon: Routledge.

Bulkeley, H., Castán Broto, V., Edwards, G., & Luque-Ayala, A. (2015). Solar hot water and housing systems in Sao Paulo, Brazil. In H. Bulkeley, V. Castán Broto, & G. Edwards, *An Urban Politics of Climate Change: Experimentation and the Governing of Socio-technical Transitions*. Abingdon: Routledge, 157–176.

Bulkeley, H., Edwards, G., & Fuller, S. (2014). Contesting climate justice in the city: Examining politics and practice in urban climate change experiments. *Global Environmental Change*, 25(1): 31–40.

Caprotti, F., & Cowley, R. (2017). Interrogating urban experiments. *Urban Geography*, 38 (9):1441–1450.

Castán Broto, V., & Bulkeley, H. (2013). A survey of urban climate change experiments in 100 cities. *Global Environmental Change*, 23(1): 92–102.

Edwards, G., & Bulkeley, H. (2018). Heterotopia and the urban politics of climate change experimentation. *Society and Space*, 36(2): 350–369.

Fünfgeld, H. (2015). Facilitating local climate change adaptation through transnational municipal networks. *Current Opinion in Environmental Sustainability*, 12(1): 67–73.

Geoklock e Ekos Brasil. (2012). Inventário de Emissões e Remoções Antrópicas de Gases de Efeito Estufa do Município de São Paulo para o período de 2003–2009. São Paulo.

Gordon, D. (2013). Between local innovation and global impact: Cities, networks, and the governance of climate change. *Canadian Foreign Policy Journal*, 19(3): 288–307.

Gordon, D., & Acuto, M. (2015). If cities are the solution, what are the problems? The promise and perils of urban climate leadership. In C. Johnson, N. Toly, & H. Schroeder (eds.), *The Urban Climate Challenge: Rethinking the Role of Cities in the Global Climate Regime*. New York: Routledge, 63–81.

Granziera, M. L. M., & Rei, F. C., eds. (2015). *O Futuro do Regime Internacional das Mudanças Climáticas – Aspectos Jurídicos e Institucionais*. Santos: Editalivros – SBDIMA.

Greenpeace. (2005). *Tolerância zero: chega de madeira ilegal*. São Paulo: Greenpeace Brasil.

Harvey, D. (1993). Tackling urban CO_2 emissions in Toronto. *Environment*, 35(7):16–20, 38–44.

Heinrichs, D., Krellenberg, K., & Fragkias, M. (2013). Urban responses to climate change: theories and governance practice in cities of the Global South. *International Journal of Urban and Regional Research*, 37(6): 1865–1878.

Hickmann, T. (2017). "The Reconfiguration of Authority in Global Climate Governance." International Studies Review 19(3): 430–451.

Hickmann, T., Fuhr, H., Höhne, C., Lederer, M., & Stehle, F. (2017). Carbon governance arrangements and the nation-state: The reconfiguration of public authority in developing countries. *Public Administration and Development*, 37(5): 331–343.

Hoffmann, M. (2011). *Climate Governance at the Crossroads: Experimenting with a Global Response after Kyoto*. New York: Oxford University Press.

ICLEI – Local Governments for Sustainability. (1997). ICLEI Biennial Report 1996–1997. Toronto: ICLEI, 1997. http://archive.iclei.org/documents/Global/biennial96-97.pdf (accessed 7 January 2019).

ICLEI – Local Governments for Sustainability. (2017). ICLEI Corporate Report 2016–2017. Bonn: ICLEI. http://e-lib.iclei.org/iclei-corporate-report-2016–2017/ (accessed 7 January 2019).

Jordan, A. J., Huitema, D., Hildén, M., Van Asselt, H., Rayner, T. J., & Schoenefeld, J. J. (2015). Emergence of polycentric climate governance and its future prospects. *Nature Climate Change*, 5(11): 977–982.

Kern, K., & Bulkeley, H. (2009). Cities, Europeanization and multi-level governance: Governing climate change through transnational municipal networks. *Journal of Common Market Studies*, 47(2): 309–332.

Krause, R. M. (2012). Political decision-making and the local provision of public goods: The case of municipal climate protection in the US. *Urban Studies*, 49(11): 2399–2417.

Krause, R. M., Hongtao, Y., & Feiock, R. C. (2016). Applying policy termination theory to the abandonment of climate protection initiatives by U.S. local governments. *Policy Studies Journal*, 44(2): 176–195.

Lambright, W. H., Chagnon, S. A., & Harvey, D L. D. (1996). Urban reactions to the global warming issue: Agenda setting in Toronto and Chicago. *Climatic Change*, 34: 463–478.

Lee, T. (2015). *Global Cities and Climate Change: The Translocal Relations of Environmental Governance*. London: Routledge.

Lee, T., & Koski, C. (2014). Mitigating climate change in global cities: Comparing the effects of participation in transnational municipal network on climate change policies of C40 cities. *Journal of Comparative Policy Analysis*, 16(1): 475–492.

Macedo, L. S. V. (2017). *Participação das Cidades Brasileiras na Governança Multinível das Mudanças Climáticas*. PhD thesis. Universidade de São Paulo – USP, São Paulo.

Macedo, L. S. V., Setzer, J., & Rei, F. (2016). Transnational action fostering climate protection in the City of São Paulo and beyond. *disP – The Planning Review*, 52(2): 35–44.

Martins, R. D'a., & Ferreira, L. C. (2011). Climate change action at the city level: Tales from two megacities in Brazil. *Management of Environmental Quality: An International Journal*, 22(3): 344–357.

Matschoss, K., & Repo, P. (2018). Governance experiments in climate action – empirical findings from the 28 European Union countries. *Environmental Politics*. DOI:10.1080/09644016.2018.1443743.

Meckling, J., Kelsey, N., Biber, E., & Zysman, J. (2015). Winning coalitions for climate policy. *Science*, 349(6253): 1170–1171.

Pitt, D. (2010). The impact of internal and external characteristics on the adoption of climate mitigation policies by US municipalities. *Environment and Planning C: Government and Policy*, 28(1): 851–871.

Rashidi, K., & Patt, A. (2018). Subsistence over symbolism: The role of transnational municipal networks on cities' climate policy innovation and adoption. *Mitigation and Adaptation Strategies for Global Change*, 23(1): 507–523.

Raven, R., Sengers, F., Spaeth, P., Xie, L., Cheshmehzangi, A., & de Jong, M. (2017). Urban experimentation and institutional arrangements. *European Planning Studies*. DOI:10.1080/09654313.2017.1393047.

Salomon, M. (2011). Paradiplomacy in the developing world. In M. Amen, N. J. Toly, P. L. McCarney, & K. Segbers (eds.), *Cities and Global Governance: New Sites for International Relations*. Farnham: Ashgate, 45–68.

Schroeder, H., & Bulkeley, H. (2009). Global cities and the governance of climate change: What is the role of law in cities? *Fordham Urban Law Journal*, 36(2): 314–360.

SEEG (Sistema de Estimativas de Emissões e Remoções de Gases de Efeito Estufa). (2017). Relatório SEEG Brasil 2017 – Documento de análise. São Paulo: Observatório do Clima.

Setzer, J. (2009). Subnational and transnational climate change governance: Evidence from the state and city of São Paulo, Brazil. In 5th Urban Research Symposium. http://siteresources.worldbank.org/INTURBANDEVELOPMENT/Resources/336387-1256566800920/6505269-1268260567624/Setzer.pdf (accessed 7 January 2019).

Setzer, J. (2013). *Environmental paradiplomacy: The engagement of the Brazilian state of São Paulo in international environmental relations*. PhD thesis, London School of Economics and Political Science (LSE), London.

Setzer, J., Macedo, L. V., & Rei, F. C. (2015). Combining local and transnational action in the adoption and implementation of climate policies in the city of São Paulo. In J. Craig, N. Toly, & H. Schroeder (eds.), *The Urban Climate Challenge: Rethinking the Role of Cities in the Global Climate Regime*. London: Routledge.

Toly, N. J. (2008). Transnational municipal networks in climate politics: From global governance to global politics. *Globalizations*, 5(3): 341–356.

Vargas, M. C., & Freitas, D. (2009). Regime internacional de mudanças climáticas e cooperação descentralizada: o papel das grandes cidades nas políticas de adaptação e mitigação. In D. J. Hogan & E. Marandola Jr. (eds.), *População e mudanças climáticas. Dimensões humanas das mudanças ambientais globais*. NEPO/ Unicamp. Brasília: UNFPA.

Vigevani, T. (2006). Problemas para a Atividade Internacional das Unidades Subnacionais: estados e municípios brasileiros. *Revista Brasileira de Ciências Sociais*, 21(62): 127–139.

Voytenko, Y., McCormick, K., Evans, J., & Schliwa, G. (2016). Urban living labs for sustainability and low carbon cities in Europe: Towards a research agenda. *Journal of Cleaner Production*, 123(1): 45–54.

Yienger, J., Brown, L., & Skinner, N. (2002). Experiences of ICLEI's Cities for Climate Protection Campaign (CCP): A focus on Asia. In *Proceedings of IGES/APN Mega-City Project*. Institute for Global Environmental Strategies.

5

Making Climates through the City

LAUREN RICKARDS

Introduction

The politics of the urban environment is core to the politics of urban climate *change* and the unevenly distributed, varied types of agencies involved. Paradoxically, however, fine-grained, actual existing urban environmental, notably climatic, conditions are given relatively scant attention in most climate change discussions. Amplified by messages about the Anthropocene and the long-distant effects urbanization is having on the planet, there is a growing interest in cities as a physical, generative force. But for the most part, this physical role is understood as mediated by the greenhouse gas (GHG) emissions that cities produce directly and indirectly. As Bulkeley (2013: 229) puts it: 'Climate change is not simply happening to cities, as a suite of environmental processes and events that cities need to endure and overcome. Rather, climate change is actively being *produced* through the urban condition.'

While the planetary significance of urban-generated GHG emissions should not be under estimated, to fully comprehend the politics of climate change and the broader Anthropocene condition, it is important to appreciate the other ways in which the urban condition produces climates and environment, and how these are differentially experienced today. By taking seriously urban climate and urban physicality more broadly, we can open up important new political considerations, conceptual lenses, and research directions. Researchers and practitioners alike need to not just take the particularities of climate seriously, or to think of climate change in urban terms, but to think of cities in environmental terms, which is to say, in more relational, material, historical and political terms. Much existing discussion about urban climate change politics obfuscates the fact that urban climates are human-made not only via the feedback loop of global climate change or via the capacity to create temperature-controlled enclaves, but because all city features – built and natural – are necessarily microclimatic interventions with implications for the quality of life of those near and far. Recognizing the climatic and thus social significance of seemingly mundane decisions about urban materials, forms, and placements opens up a new horizon of urban climate politics, one that offers both an expanded sense of human agency at the same time as tying its operation to a dense assemblage of other entities.

In the existing literature and climate change efforts, climates are recognized as urbanized or urban-made in three main ways: global climate change, the deliberate production of certain climate-controlled spaces, and the direct but unintentional climatic effects of the entire city landscape. While the first two of these are prominent issues, the latter – the direct climatic effects of city environments – remains badly neglected. In this chapter, I review these three ways in which the urban makes climates before considering why actual urban climates have a low profile in much climate change literature. I end by pointing to the value of positioning climate change in broader discussions of the Earth System of the sort that Earth System Governance (ESG) scholarship does. Along the way, I outline the relevance of an assemblage and governmentality reading of agency.

Urban-Made Climate 1: Human-Induced Global Climate Change

The question of how cities shape the climate is understood primarily via their effect on global climate change. Existing work on this important topic has helped us rethink the urban in more processual, expansive terms. But it also exposes some of the conundrums in climate change governance, notably the tension between classic top down and bottom up thinking. Combined with the need to appreciate the more-than-human agencies involved, these issues underline the value of thinking about climate change in assemblage and governmentality terms.

It is now well recognized that cities produce a substantial proportion of the world's GHG emissions. This identification of the urban as a hotspot of emissions is part of a broader move to understand social and spatial discrepancies in where, what and who unwanted atmospheric GHGs are coming from. While debates about the relative contributions of different nations or regions are caught up in technical disputes about the measurements being used, one of the least disputed patterns is that urban areas are increasingly more responsible for GHG emissions than rural areas (Sethi & Puppim de Oliveira, 2015). Adding to this focus on the urban has been efforts to think about the urban not as a spatial container but in more processual terms, with economic or life cycle analyses revealing cities as the ultimate source of most of the carbon-intensive industrial processes that now stretch across the planet (Lombardi et al., 2017). Although many such industrial processes (including not just heavy industry but also industries such as tourism) leave some of their dirtiest footprints within 'rural' or even 'remote' areas, because the actors involved are generally headquartered in cities (whether state agencies or corporations within, for example, the extractive, energy, shipping or waste management industries), and because cities are also home to most of the consumers of the ultimate products and services, these processes can be justifiably considered urban-genic.

Strengthening awareness of urban areas' environmental footprints are theoretical moves to redefine 'the urban' itself to include areas outside city limits per se that are increasingly being deeply remade by urban-genic processes. Such is the extent of these urban-genic impacts that the whole of the planet is considered by some to be 'urbanized', an argument that reinforces and has been reinforced by the idea that the whole planet is now in the 'human dominated' Anthropocene epoch. In particular, Neil Brenner and colleagues have revived Henri Lefebvre's idea of 'planetary urbanization', which reframes the urban in processual, planetary terms. In this view, specific urban forms are 'temporary moments in a wider urban process' (Schmid et al., 2018: 20). What we think of as urbanization is complicated and expanded to include three interlinked physico-economic urbanization processes; condensed urbanization in which cities (broadly defined) are expanding into peri-urban zones; extended urbanization in which cities are extracting resources from and sending wastes to distant spaces; and differentiated urbanization, in which cities are continually destroyed and rebuilt from within, typically using resources sourced through the extended urbanization process (Brenner & Schmid, 2014).

An upshot of this double move – growing awareness of the centrality of urban processes in producing climate change, and growing awareness of how immense and complex urbanization is – is that the challenge of mitigating urbanization's effects on the global climate emerges as a mammoth task. It is thus no surprise that there is a growing sense that the key urbanization challenge is to 'steer' and 'navigate' it rather than, as some once imagined, to try to stop or reverse it. The parallels and direct intersection with conversations about the Anthropocene are stark. As ESG scholars such as Eva Lövbrand have argued, one of the difficulties with many explanations of the Anthropocene is that it is presented as *a* fait accompli, a fact about the present-day or future Earth that all we must accept and manage as well as possible. Like some Anthropocene narratives (e.g. the *Welcome to the Anthropocene* film shown at the opening of the UN Rio +20 Summit), the UN Habitat New Urban Agenda is premised on a story about the planet having entered an Urban Age. Centred on the (disputable) fact that more people live in cites than outside them, the Urban Age is presented as not just irreversible and teleological, but as some kind of an achievement, reflecting long-standing representations of cities as icons of human ingenuity, sociality and progress (Rickards et al., 2016).

Others interpret both the Anthropocene and planetary urbanization in far darker terms as expressions of capitalism's destructive logic (e.g. Arboleda, 2015; Moore, 2015). Either way, the point here is that the linking of global climate change to global processes of urbanization is contributing to the framing of climate change as a matter of massive systems and epic processes (the Anthropocene, human social evolution, capitalism). The upshot is that urbanization emerges as a seemingly structural, universal, inexorable process in which responsibility is obtuse, and options for driving positive change are unclear, contributing to the paralysis that has afflicted much climate change governance to date.

In contrast to this structural reading of urbanization are efforts to understand it as an outcome of individual, more local-scale actors. Especially when seen as a site-by-site change process – for example, what Schmid et al. (2018) call 'plotting urbanism' – urbanization is arguably a matter of individuals and organizations living in and (re)making the urban fabric in certain ways. Besides daily acts of consumption and their long-distance effects, this includes the periodic upgrading of built facilities required because urban materials are constantly, inevitably decomposing, especially under a changing climate (Andrady et al., 2015), and pursued for reasons of social or economic gain. With social and economic processes and goals being a function of larger structural factors, however, efforts to address climate change by only intervening in individuals' daily remaking of cities without addressing these broader processes, remain inadequate.

Thus both structure and agent approaches to reducing climate change via intervening in urbanization reveal only limited opportunities for positive change. One possible route out of the structure–agency echo chamber is assemblage theory. As discussed by numerous authors, assemblage thinking is a diverse field of thought that draws some of its inspiration from complex systems theory (Dittmer, 2014). Its recent popularity reflects the fact that it provides a means of integrating the powerful effects of both material and discursive elements of the world. It also allows us to side step the conventional nesting of topographical scales (local to global) to appreciate more topological relationships and coherences, at the same time as recognizing the independent reality, contingent role and dynamic character of the entities involved (Dittmer, 2017). At a given moment, an assemblage may or may not be what Manuel DeLanda (2006) calls highly territorialized (coherent, strong and distinct) or highly coded (recognizable, laden with meaning). If we think of urbanization as an assemblage it is clear that it is not just expanding and refracting spatially, but is becoming more strongly coded within climate change discussions.

A particular advantage of assemblage thinking for examining and intervening in the production of human-induced global climate change via urbanization is that, like Actor Network Theory, it is attentive to diverse material components, whether they be human bodies, bricks, bushland, or broad more-than-human processes such as solar radiation. It thus helps highlight that urbanization is a thoroughly physical process. In particular, assemblage thinking recognizes not just that 'stuff is linked' but that it is highly significant how entities are 'oriented (physically or psychologically) towards each other' (DeLanda, 2006: 12). In these ways, assemblage thinking helps detail what agency involves, by suggesting among other things that it involves the capacity to reorient towards the world. It also expands the scope of what is recognized as political by helping map the entities, relations, forces, and orientations involved. In a sense, the aforementioned planetary urbanization thesis can be interpreted as an expanded assemblage-like reading of urbanization that helps map how processes at different sites and scales are part of a related, coded phenomenon in which distant sites are oriented towards cities.

There is a close resonance between the relational reading of power in assemblage thinking and the Foucauldian notion of governmentality that various ESG scholars have used to productively interpret dominant approaches to urban climate change governance. With Foucault's idea of apparatus (*dispositif*) representing a specific type of assemblage, governmentality is in a sense about governing through assemblages of human and more-than-human elements. It arose in the seventeenth century as the target of governance expanded from individual subjects to 'the milieu' that individuals collectively exist within and coproduce. A precursor to the idea of 'the environment', the milieu refers to the 'intersection between a multiplicity of living individuals working and coexisting with each other in a set of material elements that act on them and on which they act in turn' (Foucault, 2009: 21). It is significant to note that the question of urban climates is more than just a site to which governmentality theory can be applied. Rather, urbanization and climate are foundational 'milieu' challenges. Foucault discussed the problem of 'the town' as one of the issues that gave rise to governmentality, posing as the town did the question of how to manage circulations of people and things (Foucault, 2009). Although conventionally positioned in social discourse as (relatively) natural, climate is also presented by Foucault as one of the first targets of 'milieu thinking', not just because it was seen (e.g. by European colonial settlers) as a determinant of human character, but also because it was seen as somewhat manipulatable. Foucault presents French demographer Jean Baptiste Moheau as 'the first great theorist of what we could call biopolitics' for his assertions that 'it is up to the government to change the air temperature and to improve the climate … [to create] a new climate' in order 'to govern the physical and moral existence of their subjects' and try to 'give whatever hue we wish to morality and the national spirit' (Moheau, 1778, in Foucault, 2009: 22–23).

Urban-Made Climate 2: Enclosed Climate-Controlled Spaces

The question of air temperature brings us to the second way in which climates are commonly recognized as made in and by the urban: the deliberate, literal production of desirable atmospheres in certain spaces through the use of design, architecture, and technology (notably mechanical air-conditioning). As Foucault (2000: 148) put it, government became a matter of 'controlling circulation'. Not the circulation of individuals but of things and elements, mainly water and air'. Control of air circulation has allowed the creation of indoor temperature- and humidity-controlled microclimates. Such 'thermal enclosures' (Rickards & Oppermann, 2018: 4) are expanding in scale to such an extent – from the 'capsule' of the building to the 'dome' of the mall (Marvin, 2015) – that, combined with air-conditioned transportation, a growing number of urban dwellers live within a web of artificial 'ecological enclaves' (Graham & Marvin, 2001). Consistent with the rise of 'volumetric urbanism' (Marvin, 2015), the indoors is a spatial frontier expanding throughout and through cities as more and more volumetric space is enclosed. But the opportunity to live within controlled microclimates is unevenly distributed socially, a classed privilege consistent with the gentrifying effects of much differential urbanism. In the context of the increasingly pathological state of the 'wild climate' outdoors, the politics of air and its differential conditioning is emerging as a major axis of urban climate change politics.

The agency of air in cross-scale assemblages of domestic, professional, national and transnational spaces and bodies has long been implicitly recognized as culturally and economically important. In the era of Imperial expansion, tropical conditions were conflated by European settler colonialists such as Moheau with under-development, soluble only through a combination of clever architecture, design, and personal hygiene (Rickards & Oppermann, 2018). In the post-WWII 'Great Acceleration' of the Anthropocene, the idea of 'thermal comfort' – the assertion that there is a universal temperature that all humans perform best at – was introduced by the US air-conditioning industry as part of a calculated effort to claim air space as a profitable frontier (Cooper, 2002). They successfully convinced employers and government to recognize – in quintessential governmental fashion – that individuals could be made more industrious if the air conditions around them were kept within a narrow band (Ackermann, 2013). With the subsequent institutionalization of thermal comfort standards across the world, at the same time as external temperatures are being rendered more problematic by climate change, climate control is increasingly one of 'the practices through which we govern, and are governed, in our everyday lives' (Lövbrand & Stripple, 2014: 111). As a growing number of critics point out, however, artificially controlled urban microclimates are problematic not just as a governmental technique in aid of capitalism and imperialism, but because they are generally highly energy- and emissions-intensive, thus exacerbating the global climate change that is perpetuating their production.

Urban-Made Climate 3: The Climatic Effects of Urban Landscapes

The preceding discussion makes clear that urban climates are on the agenda as an expression of and particular response to human-induced climate change. But there is still a basic aspect of the urban–climate relationship that remains under-examined: the way the urban environment itself inevitably coproduces atmospheric conditions, weather, and climate. Above and beyond the effect of urbanization on global climate change, the deliberate cultivation of select microclimates within cities, and even the effects of associated heat-emitting machinic processes, urban climates are shaped by the interacting chemical, physical, and, in some cases, biological qualities of different aspects of the urban environment and how they are arranged in three-dimensional space. As Webb (2017: 69) puts it:

> Whether in desert or tundra, coastline or valley, each human settlement has a distinctive climate, modified by the form and mass of its buildings and by the configuration of its streets and open spaces. Its design markedly affects its temperature, humidity, air movement and even precipitation . . .

Some important steps have been taken towards thinking of cities as not situated within the environment but as themselves *an environment*. Most obviously, the turn to green infrastructure represents an effort to start thinking of cities in both infrastructural and environmental terms, with the two concepts themselves converging. However, although conceptually and practically important, efforts to insert more vegetation and water into cities in order to create cooler microclimates, among other things, stop short of acknowledging the fundamentally climate-genic character of each and every bit of cities. Above and beyond their planned and infrastructural parts, cities are an environment coproduced by the relations between physical components of various densities and other qualities. One component is what we call the atmosphere, which comes with associated weather and climatic characteristics that affect, in turn, the form and functioning of other, denser parts, including human bodies.

To date, the main vehicle for conceiving of the coproduction of cities and local climates is the idea of the Urban Heat Island (UHI) effect. As a strongly coded assemblage of trapped heat, materials, models, and calculations, the UHI effect draws much-needed attention to the often-unintentional climatic effects of urban planning and design, helping shift the focus from ‘canalized’ circulations of people and things of the sort that preoccupied Foucault (see Foucault, 2000), and now infrastructure scholars, towards more basic questions about mass, radiation, orientation, surfaces and the social politics of materials. It usefully emphasizes that 'The built environment, including buildings and roadways that absorb sunlight and re-radiate heat, combined with less vegetative cover to provide shade and hold cooling moisture, all contribute to cities being warmer and susceptible to dangerous heat events' (Corburn, 2009: 416).

(Harlan & Ruddell, 2011: 127) argue that, irrespective of GHG effects:

> 'The transformation of native landscapes into dense urban settlements of heat-retaining impervious surfaces and building materials that inhibit night-time cooling is the most significant anthropogenic driver of urban climate change in cities around the world.'

Rising awareness of how the production of urban spaces has unintentionally inhibited night-time cooling in cities helps bring into view the broader point that

> 'every intervention in the urban tissue is in fact an intervention in the urban climate, and . . . the urban climate is thus largely 'designed' by those who make the city' (Lenzholzer, 2015: 14).

There are at least two benefits to this way of thinking about cities. First, it brings to the fore the largely unconscious ways that urban environments are actively produced. This highlights, in turn, the uneven distribution of different environmental qualities and the justice implications of the choices or at least processes that give rise to such patterns. That is, it deepens our grasp of the politics of urban climate change by underlining how constructed and political urban climates are *per se*. In doing so, it contributes to 'a more thoroughly political analysis of the urban climate governance problematic' (Bulkeley & Betsill, 2013: 151). Second, like the Anthropocene that ESG scholars have helped scrutinize, it alters and deepens our sense of human agency in two ways. On the one hand, it emphasizes the extent of certain human terraforming – including climate-forming – capacities. On the other hand, it underlines that not only is this world-making agency only partially controlled, it is only partially controll*able*, emerging as it does out of complex interactions and distributed power of the material-discursive assemblages we are part of. In other words, it illuminates the paradoxes at the heart of the human condition in the Anthropocene, offering a vital laboratory for explorations of how this new awareness may be directed at improved outcomes.

Despite the particular and general benefits it offers, thinking of cities as environment is currently not widespread. Nor is a more specific focus on intervening in outdoor urban climates. Lenzholzer (2015) suggests that few urban managers, for example, consciously design or govern for specific urban climatic conditions. As a result, the deeply uneven climatic qualities of cities remain largely under-examined and unaddressed. Arguably, multiple aspects of the existing climate change literature contribute to this obfuscation of the local as well as global ways in which urban climates are produced. We turn now to consider this literature and the ways it backgrounds actual urban climates.

Clouding Actual Urban Climates in Climate Change Literature

As critical ESG scholars have highlighted about the Anthropocene, all discourses foreground some matters and concerns relative to others. In the dominant social science literature on urban climate change, including that by ESG scholars, the following nine topics have received notable attention: governance, the global climate, the urban, futures, climate *change*, carbon, social vulnerability, adaptation, and geoengineering. While resultant scholarship and practice is of value, collectively it has inadvertently side-lined the material particularities of current city climates, and with them, their historical, cultural, uneven and modifiable character (Table 5.1). Although work on each topic tends to compensate for the blind spots of others to some degree, as a whole they are relatively silent on how urban climates are being produced today, why and to what effect. We look at each of the nine topics in turn below.

One of the critiques that ESG scholars and others have had of the core scientific discourse about the Anthropocene is that when it comes to our first topic here – governance – the scientific narratives need to be complemented and/or challenged by scholarship on the immense social issues involved. That includes work from a governmentality lens that focuses attention on the *mentalities* of governing. A risk of this move, however, is a relative neglect of the physical particularities and complexities. As mentioned above, some governmentality studies and related developments in assemblage, Actor Network Theory and infrastructure studies, have started to counter an emphasis on discursive factors with an emphasis on the agency of material entities, from 'recalcitrant' more-than-human actors in scientific experiments such as microbes to hard copy reports, digital interfaces, meeting rooms, electricity lines and plants. Yet, the more-than-human elements considered tend to be 'objects' within an otherwise invisible physical matrix. Although this 'thing bias' is beginning to be challenged by the reconceptualization of (urban) nature such as soil or urban forests as bio-infrastructure (e.g. Puig de la Bellacasa, 2014), the focus remains on materially tangible factors, not the general milieu. Environmental justice research usefully considers atmospheric conditions, but generally only to the extent they are considered pathological. That is, it focuses on pollutants, not 'normal conditions', and has only recently begun to consider climatic aspects such as heat (e.g. Mitchell & Chakraborty, 2015). Overall, the diverse physical qualities of different urban environments such as the radiation characteristics of different surfaces, their continual daily production, their uneven

Table 5.1 *Nine topics shaping research on urban climate futures, aspects of the issue they foreground and aspects relevant to actual urban climates that they background*

Topic	Foregrounds	Backgrounds
Governance	Social (science) not scientific/ physical considerations	Physical conditions and their historical production
The global climate	Global scale; climate as statistical, probabilistic aggregate	Local scale, observed climate; urban climatology; lived experience of climate
The urban	Urban as generic category, framed in various ways	Particular, varied, changing, material cityscapes
Futures	What lies 'ahead' in time; changes	The present day and the past; continuities
Climate *change*	What is different; exceptions not the norm (e.g. climatic extremes)	Past climates; climatic continuities; everyday statistically normal climates
Carbon	Effects of changing GHG concentrations on the global climate and need to govern them	Need to govern climate *per se;* other ways urban climates are created
Social vulnerability	Social and economic aspects of contextual vulnerability to climate change risks	Physical vulnerabilities including existing climatic stresses
Adaptation	Actions to moderate climate risks or reduce vulnerability to them	How to moderate not only vulnerability to climate, but unevenly distributed climatic conditions
Geoengineering	Large-scale deliberate, experimental interventions to reduce average global temperature	Efforts to modify climate more locally; weather modification techniques beyond Cold War technologies

distribution, their social effects and their actual agencies in different assemblages remain under-examined.

Second, urban climates specifically are obscured by the dominant coding of climate as global. An area of research that studies urban climates explicitly is urban climatology. But as Hebbert and Jankovic (2013) and Webb (2017) note, urban climatology has not been widely adopted in urban planning, leading to very few cities being designed in a way cognizant of the climates being created. One reason for this institutional marginalization is that urban climatology was overshadowed by the focus on global climate and global climatology, leading to a sense that climate is exogenous and given, not endogenous and open to change. By occluding smaller scale differences in climate, the focus on a singular global climate disguises the fact that it was colonial era encounters with regional climatic differences that generated the very idea that something called *climate* exists (Mahony & Endfield, 2018). As mentioned earlier, the colonial era representation of The Tropics as 'abnormal' contributed to the idea that different climates exist, some are better than others, and hot ones should be ameliorated (Chang, 2016; Hulme, 2008). Thus almost since its first emergence as an idea, climate was recognized as modifiable and shaped by urban settlements. These crucial historical linkages have been papered out in the contemporary discussions of urban climates as simply subsets of the global climate and efforts to govern it. Adding them back in promises to strengthen our understanding of the character of contemporary governmentalities entwined with urban climate futures.

That brings us to the third topic: the urban. Whether imagined through a governmental lens focused on population, a section of sovereign territory, or as a more modern network hub or scale, the particular physical characteristics of specific city spaces tends to be overlooked by a focus on socio-political factors and an epistemological inclination towards generality. The urban scale of analysis and focus on built forms also tends to imagine urban life as primarily lived indoors, whether inside buildings or forms of transport. People's daily encounters with weather slip between the cracks even as these encounters strongly shape people's everyday practices and lives. Helping counter this neglect of outdoor climatic conditions in cities is work on the UHI effect as mentioned earlier. But it can create a static and deterministic picture of the dynamics involved, painting over the particularities and politics of the use of specific materials, masses, spatial configurations and energetic processes. Its focus on heat distracts from other ways urban settings interact with atmosphere, weather, and climate. As sufferers of Thunderstorm Asthma know, for example, the interaction of urban greenery, certain seasonal climatic conditions and warm, stormy weather can alter the composition of the urban atmosphere in fatal ways (Demain, 2018). As a relative concept, the UHI effect problematizes urban temperatures because they are higher than surrounding areas but this can detract from the important point that all landscapes co-create microclimates and that in any place the latter may create problems and injustices for those involved.

Lived experiences are further obscured by a focus in much climate change governance on the fourth topic, the future. As others have noted in relation to climate change messaging based on projections of future climates, the temporal framing of climate change as a matter of the future distracts attention from people's existing challenges, including climatic ones, as Houston (2013) discusses in relation to the Anthropocene. More specifically, this modernist preoccupation with the future interprets the present in terms of the future so that only certain things are seen as relevant (Edwards & Bulkeley, 2017).

That brings us to the fifth topic, climate *change*, which adds to this funnelling by directing attention at how emerging climates *differ* from existing ones. In this imaginary, troubles are located in the new not the old. Thus, interest in the UHI effect, for example, has grown because it is being exacerbated by climate change. The focus is not on chronic thermal stress as much as increases in temperature extremes and acute heat stress ('heat as emergency' (Bolitho & Miller, 2016). The scale of work also reflects the climate change influence, with a reliance on downscaled global climate science rather than local urban climatology (e.g. Corburn, 2009) adding to a sense that the UHI effect is a generic urban characteristic rather than a diverse outcome of the specific ways different city designs generate, alter, and unevenly distribute heat at a neighbourhood and building scale. For example, a Dutch study found that during hot periods people's thermal comfort varied greatly according to the shape of the outdoor space they were in, with courtyards found to be most comfortable (Taleghani et al., 2015).

Related to the focus on global climate change is a focus on its cause: GHGs, notably carbon dioxide. Much climate change governance literature is understandably focused on reducing these emissions, including influential work by ESG scholars using a governmentality lens (e.g. Bulkeley et al., 2016; Lövbrand & Stripple, 2014; Stripple & Bulkeley, 2015). Some work looks explicitly at the relationship between emissions and urban form, leading to interest in creating more 'compact' cities, for example. Such a focus on densification, however, has arguably distracted attention from the fact that such urban forms are often hotter: the 'so called "paradox of the compact city"' (Koch et al., 2018: 31; Pearsall, 2017). Although 'smart urban planning approaches can ... lead to cool *and* compact urban areas' (Koch et al., 2018: 31), they require a detailed understanding of the role of urban form in city temperatures. More than just inserting cooling technologies such as trees or green roofs, this requires reassessing existing urban forms, including their material characteristics, surfaces, colour, spatial layout and orientations. These more localized spatial character and effects are generally considered secondarily, despite the latter shaping their relative local influence.

The seventh topic is social vulnerability. One of the key debates around climate change futures has been the distribution of climate change impacts. Indicative of the science-first evolution of climate change thinking (repeated with the Anthropocene), the primary focus has been *climatic* impacts. To counter this overly simplistic framing of climate change, over the last two decades social scientists such as Neil Adger and Karen O'Brien have emphasized that impacts are coproduced by people's existing contextual vulnerability. Being motivated by a desire to challenge the scientistic interpretation of climate change impacts, this work has turned attention away from climatic factors to socioeconomic factors, as indicated by the indicators of vulnerability now used in many climate change vulnerability assessments (e.g. Preston et al., 2011). An unintended outcome of this has been a downplaying of the intimate role of physical climate in people's everyday lives and vulnerability profiles, not just in terms of hazardous weather experiences, but their capacity to, for example, regularly exercise outdoors, use public transport, or grow their own vegetables. Where climate is considered it is generally framed as a *risk*; i.e. it is placed in the future. For example, a recent analysis of New York, Los Angeles, and Chicago concluded that there is a 'consistent and significant statistical association between lower socioeconomic and minority status and greater urban heat *risk*' (Mitchell & Chakraborty, 2015: 115005, italics added). While a useful insight, this and similar studies obscure existing climatic stress, thus narrowing the scope of urban climate change politics discussed.

Impacts bring us to the eighth topic, which is adaptation. Critical literature on adaptation points to the predominantly incremental and individualistic approach that dominates adaptation responses to date (e.g. Bassett & Fogelman, 2013; Pelling, 2011). Adaptation recommendations around heat stress, for example, are centred on technologies, behaviour, and indoor often private spaces, namely calling for people to remain enclosed within air-conditioned enclaves during hot periods. This externalizes the common spaces of the city on which some rely more than others (Khalil et al., 2018) and raises serious energy justice questions due to the electricity or fuel involved being less relatively affordable for residents or commuters already under financial stress (Nicholls & Strengers, 2017). Not only do resultant GHG emissions (from refrigerants as well as energy sources) then worsen climate change at the global level, the waste heat from air conditioners and other machines in the city adds considerably to urban heating and the level of overall heat people experience (Harlan et al., 2007). Who can access machine-free areas, mid-day shade, afternoon breeze and cooling rain, and who has to suffer the un-ameliorated effects of radiated heat, metallic glare and hot exhaust fumes, are revealed as political, environmental justice matters, but are yet to be examined in current adaptation discussions, that are focused instead of access to the 'cabin ecologies' of the contemporary air-conditioned city (Marvin & Hodson, 2016).

The final topic is geoengineering. Critiques of calls to engineer the Anthropocene Earth to improve human outcomes, including by those writing from an ESG perspective (e.g. Talberg et al., 2018), rightly point out how hubristic and risky such notions are. In particular, global scale initiatives and solar radiation management have been dismissed as outlandish and dangerous. Yet, in a sense geoengineering's focus on the general fact of how humans can shape, and already have shaped, their environmental conditions, including via efforts to alter solar radiation relations, is one that is needed, at least at scales below the planet. Some literature on geoengineering usefully points to historical precedents of the 'mentality of climate control' (Buck, 2012: 257) and its application at sub-global scales. But its focus on a particular military-industrial history of weather control detracts from longer-standing, more diffuse and more benign efforts to modify weather, including those using urban form. Thus important conversations about the agentic role of urban design and urbanization in current as well as future climate-human assemblages are cut short.

Notwithstanding these multiple obfuscations of the urban climate and broader urban environment in much climate change literature, important openings for thinking more intently about city-made climates are opening up, particularly in the ESG scholarship. In the following, final section, I review a few of these openings and summarize some of their insights.

Drawing from Earth System Science to Expand Urban–Climate Politics

ESG scholars Lövbrand et al. (2015: 217) call for a new 'generation of Anthropocene scholarship' in which social science helps drive a 're-politicization' of the processes that have brought the Anthropocene assemblage – as a discursively coded set of relations between socio-material phenomena – into being. While deeply critical of aspects of the Anthropocene thesis, these and other ESG scholars do not dismiss the underpinning sciences out of hand. For example, Uhrqvist and Lövbrand (2014) empathetically describe the contradictions, struggles, ambiguities and forced intellectual humility inherent to Earth system scientists' efforts to know the Earth as a single system. In this section, I suggest there is merit in engaging further with some of the specific suggestions of ESS above and beyond the proposal of an Anthropocene epoch. In contrast to the nihilism of stratigraphy (see Rickards, 2015), thinking in Earth System terms offers valuable insights about agency and politics, including those involved in the deep inter-relations between urban settings and climate. Indeed, positioning climate change within broader discussions of the Earth System and its governance is vital for getting to grips with the full extent and dimensions of the climate change problem.

Two aspects of an Earth System lens help illustrate this. Both represent well-recognized ideas about systems thinking as a joining of dots, but also push further towards a greater engagement with the *specifics* of the systems involved. First, thinking in terms of the relations between Earth subsystems helps underline that the atmosphere – the classic object of climate change work – is not separate to the land. As a specific volumetric space the latter is further presented as home to interactions between other subsystems such as the lithosphere, biosphere, hydrosphere and technosphere. By analysing their constant four-dimensional interaction, ESS encourages a processual, historical, relational view of the world of the sort taken up in assemblage thinking. A particular focus on the movements of gases, including terrestrial carbon sequestration, usefully encourages awareness of the broad, diffuse relations shaping climate. Together these insights frame the urban atmosphere as inseparable from the urban landscape and the urban climate as partially an emergent effect of their interaction.

Second, ESS thinks of the planet in thermodynamic terms in which qualitative categorical differences – between, for example, a road, rooftop, grassland, soil, human body and air mass – give way to thermodynamic measures that perceive such things as substances across which other patterns of difference are evident (e.g. relative penetrability to certain radiation wavelengths). Combined with the way ESS and especially solar radiation management draws attention to the Earth's orientation to the sun, this focus on material differences, interactions, orientations and their consequences for different lived experience open up a new field of vision and a new horizon for urban climate politics.

Together these two aspects of an ESS lens illustrate the way ESS thinking holds progressive potential above and beyond that acknowledged in ESG scholarship to date. As discussed, existing work by ESG scholars includes research that underlines the value of adopting a Foucauldian perspective on power as diffuse and distributed. Perhaps via assemblage thinking and its emphasis on more-than-human agencies and complexity, one opportunity for future research in this area is to engage with the specifics of ESS more deeply. Among other things, this promises to expand our understanding of the urban–climate relation and thus the politics of urban climate change. More than a matter of emissions or access to climate-controlled enclaves, the latter needs to be rooted in an understanding that climate is not just something that a city sits within or has happen to it, but is instead partially *of the city.* Global human-induced climate change is thus not about the sudden arrival of climate in the city, but is rather a twisting of and enlargement of an *already anthropogenic climate*. Such a realization offers an expanded view of human agency even as it underscores that any such agency is necessarily about manoeuvering and reorienting within thick webs of relations. Intellectually and practically, engaging in this way with cities as environments is essential if we are to learn to act in ways that makes our uneven, collective terraforming of the Earth explicit, transparent and just, 'rather than an afterthought' (Dalby, 2016: 37).

References

Ackermann, M. (2013). *Cool Comfort: America's Romance with Air-Conditioning.* Washington, DC: Smithsonian Institution.

Andrady, A., Torikai, A., Redhwi, H., et al. (2015). Consequences of stratospheric ozone depletion and climate change on the use of materials. *Photochemical & Photobiological Sciences*, 14(1): 170–184.

Arboleda, M. (2015). In the nature of the non-city: Expanded infrastructural networks and the political ecology of planetary urbanisation. *Antipode*: n/a-n/a.

Bassett, T. J., & Fogelman, C. (2013). Déjà vu or something new? The adaptation concept in the climate change literature. *Geoforum*, 48: 42–53.

Bolitho, A., & Miller, F. (2016). Heat as emergency, heat as chronic stress: Policy and institutional responses to vulnerability to extreme heat. *Local Environment*, 1–17.

Brenner, N., & Schmid, C. (2014). The 'urban age' in question. *International Journal of Urban and Regional Research*, 38(3): 731–755.

Buck, H. (2012). Geoengineering: Re-making climate for profit or humanitarian intervention? *Development and Change*, 43(1): 253–270.

Bulkeley, H. (2013). *Cities and Climate Change*. London: Routledge.

Bulkeley, H., & Betsill, M. M. (2013). Revisiting the urban politics of climate change. *Environmental Politics*, 22: 136–154.

Bulkeley, H., Powells, G., & Bell, S. (2016). Smart grids and the constitution of solar electricity conduct. *Environment and Planning A*, 48(1): 7–23.

Chang, J-H. (2016). *A Genealogy of Tropical Architecture: Colonial Networks, Nature and Technoscience*. London: Routledge.

Cooper, G. (2002). *Air-Conditioning America: Engineers and the Controlled Environment, 1900–1960*. Baltimore, MD: Johns Hopkins University Press.

Corburn, J. (2009). Cities, climate change and urban heat island mitigation: Localising global environmental science. *Urban Studies*, 46(2): 413–427.

Dalby, S. (2016). Climate security in the Anthropocene. In P. Wapner & H. Elver (eds.), *Reimagining Climate Change*. London: Earthscan, 29–48.

DeLanda, M. (2006). *A New Philosophy of Society: Assemblage Theory and Social Complexity*. London: A&C Black.

Demain, J. G. (2018). Climate change and the impact on respiratory and allergic disease. *Current Allergy and Asthma Reports*, 18(4): 22.

Dittmer, J. (2014). Geopolitical assemblages and complexity. *Progress in Human Geography*, 38: 385–401.

Dittmer, J. (2017). *Diplomatic Material: Affect, Assemblage, and Foreign Policy*. Durham, NC: Duke University Press.

Edwards, G. A., & Bulkeley, H. (2017). Heterotopia and the urban politics of climate change experimentation. *Environment and Planning D: Society and Space*. 0263775817747885.

Foucault, M. (2000). The birth of social medicine. In J. Faubion (ed.), *Power: Essential Works of Foucault 1954–1984*. London: Allen Lane, 134–156.

Foucault, M. (2009). *Security, Territory, Population: Lectures at the Collège de France 1977–1978*. London: Macmillan.

Graham, S., & Marvin, S. (2001). *Splintering Urbanism: Networked Infrastructures, Technological Mobilities and the Urban Condition*. London: Routledge.

Harlan, S. L., & Ruddell, D. M. (2011). Climate change and health in cities: Impacts of heat and air pollution and potential co-benefits from mitigation and adaptation. *Current Opinion in Environmental Sustainability*, 3(3): 126–134.

Harlan, S. L., Brazel, A. J., Darrel Jenerette, G., et al. (2007). In the shade of affluence: The inequitable distribution of the urban heat island. In Robert C. Wilkinson & William R. Freudenburg (eds.), *Equity and the Environment*. Research in Social Problems and Public Policy, Vol. 15. Bingley: Emerald Group Publishing Limited, 173–202.

Hebbert, M., & Jankovic, V. (2013). Cities and climate change: The precedents and why they matter. *Urban Studies*, 50(7): 1332–1347.

Houston, D. (2013). Crisis is where we live: Environmental justice for the Anthropocene. *Globalizations*, 10(3): 439–450.
Hulme, M. (2008). The conquering of climate: Discourses of fear and their dissolution. *Geographical Journal*, 174(1): 5–16.
Khalil, H., Ibrahim, A., Elgendy, N., et al. (2018). Could/should improving the urban climate in informal areas of fast-growing cities be an integral part of upgrading processes? Cairo case. *Urban Climate*, 24: 63–79.
Koch, F., Bilke, L., Helbig, C., et al. (2018). Compact or cool? The impact of brownfield redevelopment on inner-city micro climate. *Sustainable Cities and Society*, 38: 31–41.
Lenzholzer, S. (2015). *Weather in the City-How Design Shapes the Urban Climate.* Rotterdam: Nai 010 Uitgevers/Publishers.
Lombardi, M., Laiola, E., Tricase, C., et al. (2017). Assessing the urban carbon footprint: An overview. *Environmental Impact Assessment Review*, 66: 43–52.
Lövbrand, E., & Stripple, J. (2014). Governmentality. In C. Death (ed.), *Critical Environmental Politics*. London: Routledge, 111–120.
Lövbrand, E., Beck, S., Chilvers, J., et al. (2015). Who speaks for the future of Earth? How critical social science can extend the conversation on the Anthropocene. *Global Environmental Change*, 32: 211–218.
Mahony, M., & Endfield, G. (2018). Climate and Colonialism. *Wiley Interdisciplinary Reviews: Climate Change*. e510-n/a.
Marvin, S. (2015). Volumetric urbanism. In O. Coutard & J. Rutherford (eds.), *Beyond the Networked City: Infrastructure Reconfigurations and Urban Change in the North and South*. London: Routledge, 227.
Marvin, S., & Hodson, M. (2016). Cabin ecologies. In J. Evans, A. Karvonen, & R. Raven (eds.), *The Experimental City*. London: Routledge, 88–96.
Mitchell, B. C., & Chakraborty, J. (2015). Landscapes of thermal inequity: Disproportionate exposure to urban heat in the three largest US cities. *Environmental Research Letters*, 10(11): 115005.
Moore, J. W. (2015). *Capitalism in the Web of Life: Ecology and the Accumulation of Capital.* London: Verso Books.
Nicholls, L., & Strengers, Y. (2017). Rising Household Energy and Water Bills: Case Studies of Health, Wellbeing and Financial Impacts. Melbourne: Centre for Urban Research, RMIT University on behalf of the Victorian Council of Social Service.
Pearsall, H. (2017). Staying cool in the compact city: Vacant land and urban heating in Philadelphia, Pennsylvania. *Applied Geography*, 79: 84–92.
Pelling, M. (2011). *Adaptation to Climate Change: From Resilience to Transformation*. London: Routledge.
Preston, B. L., Yuen, E. J., & Westaway, R. M. (2011). Putting vulnerability to climate change on the map: A review of approaches, benefits, and risks. *Sustainability Science*, 6(2): 177–202.
Puig de la Bellacasa, M. (2014). Encountering bioinfrastructure: Ecological struggles and the sciences of soil. *Social Epistemology*, 28(1): 26–40.
Rickards, L. (2015). Metaphor and the Anthropocene: Presenting humans as a geological force. *Geographical Research*, 53(3): 280–287.
Rickards, L., Gleeson, B., Boyle, M., & O'Callaghan, C. (2016). Urban studies after the age of the city. *Urban Studies*, 53: 1523–1541.
Rickards, L., & Oppermann, E. (2018). Battling the tropics to settle a nation: Negotiating multiple energies, frontiers and feedback loops in Australia. *Energy Research & Social Science*, 41: 97–108.

Schmid, C., Karaman, O., Hanakata, N. C., et al. (2018). Towards a new vocabulary of urbanisation processes: A comparative approach. *Urban Studies*, 55(1): 19–52.

Sethi, M., & Puppim de Oliveira, J. (2015). From global 'North–South' to local 'Urban–Rural': A shifting paradigm in climate governance? *Urban Climate*, 14: 529–543.

Stripple, J., & Bulkeley, H. (2015). Governmentality. In K. Bäckstrand & E. Lövbrand (eds.), *Research Handbook on Climate Governance*, 49–59.

Talberg, A., Christoff, P., Thomas, S., et al. (2018). Geoengineering governance-by-default: An earth system governance perspective. *International Environmental Agreements: Politics, Law and Economics*, 18(2): 229–253.

Taleghani, M., Kleerekoper, L., Tenpierik, M., et al. (2015). Outdoor thermal comfort within five different urban forms in the Netherlands. *Building and Environment*, 83: 65–78.

Webb, B. (2017). The use of urban climatology in local climate change strategies: A comparative perspective. *International Planning Studies*, 22(2): 68–84.

6

Cross-Movement Alliances as a Novel Form of Agency to Increase Socially Just Arrangements in Urban Climate Governance

KARSTEN SCHULZ AND ANTJE BRUNS

Introduction

The important role of urbanization as a catalyst in dynamic global change processes is now widely acknowledged. Urban areas are the principal form of social organization in the twenty-first century, as well as major drivers of resource demand and environmental degradation (United Nations, 2016). The realization of socially just and sustainable urban development can thus be regarded as a precondition for the successful implementation of adaptation and mitigation actions. In view of the massive scale and impact of planetary urbanization, it can also be observed that environmental movements are increasingly aligning their ecological agendas with a focus on sustainable urban futures and social justice concerns. Unsustainable urbanization, widening social inequalities within cities and the reduction of public sector governance capacities due to privatization are recognized as crucial issues by a growing number of locally embedded 'cross-movement' alliances as well as by relatively new and more globally oriented initiatives such as the People's Climate Movement. These increasingly influential cross-movement alliances are bringing together actors from diverse environmental, rights-based, labour/unionist, and spiritual backgrounds, while making a case for innovative urban politics and deep-seated societal transformations (Carroll & Ratner, 1996; Simpson, 2015).

Drawing on examples from the metropolitan areas of Accra, Ghana, and Berlin, Germany, this chapter thus perceives the contemporary urban sphere as a highly politicized space and seeks to shed light on the emergence, roles, strategies and embeddedness of cross-movement alliances in urban climate governance. Although Berlin and Accra have roughly the same population size, it is evident that the scope and nature of the challenges which arise in both cities with regard to the required sustainability transitions differ vastly. While the city of Berlin has embraced the task of transforming its existing energy and utility systems against the background of a top-down low-carbon transition (Becker et al., 2015, 2016), the case of Accra is illustrative of the challenges that many cities need to confront while attempting to manage a necessary expansion of housing, water and energy infrastructures in a sustainable and socially inclusive manner (Amoako, 2017; Silver, 2014).

Considering the different challenges for urban climate governance in Accra and Berlin, urban political ecology (UPE) (Castán Broto & Bulkeley, 2013; Swyngedouw, 1996) is used as a unifying theoretical lens to examine how cross-movement alliances are involved in struggles over the governance of socio-technical systems such as municipal water utilities and energy infrastructures that are crucial for climate actions. While emphasizing the political nature of the nested metabolic relations that constitute the urban, UPE theory does not merely perceive unsustainable urbanization as a technical governance challenge that may be solved by means of consensual decision-making, but rather as an expression of tightly interwoven social, economic, political, cultural, material, and ecological dynamics that together 'form highly uneven urban landscapes' (Heynen, 2014: 5). Seen in this light, conflicts over the governance of urban socio-technical systems are not simply an expression of unequal urbanization trajectories, but rather symptomatic of deep-seated structural, economic and institutional factors that may stall much-needed societal transformations in the face of global climate change. Hence, UPE theory seeks to combine a view of the urban as a historically embedded formation of knowledge, power and space with a normative commitment to social and environmental justice (Rice, 2014). It also offers a critical and politically sensitive approach for examining the extent to which cross-movement alliances might empower or disempower socially just climate actions. Of course, this normative premise of UPE scholarship opens up the contentious question of why and when certain climate actions may be regarded as socially and environmentally just. Who has the agency and power to define what is just, and for whom? And how do varying normative views of contentious political issues influence the emergence as well as the leverage of cross-movement alliances in urban climate governance?

Certainly, we do not claim to provide any definitive answers to these rather generic questions about the role of cross-movement alliances in political struggles over sustainable urban futures. Instead, our aim is to shed light on the specific circumstances under which cross-movement alliances in Accra and Berlin exercise political agency, and to summarize the main lessons that can be derived from these cases. We begin our inquiry by reviewing studies on cross-movement alliances in urban climate governance that have been published over the last decade, with a particular emphasis on existing knowledge gaps and the contribution of urban political ecology to this specific topic. We continue by pinpointing the most important political factors that shape the agency of cross-movement alliances in negotiating the privatization and re-localization of urban utility systems in Accra and Berlin. The relative success of these alliances in achieving more just and equitable policy outcomes, we argue, strongly depends on their capacity to recognize and effectively exploit political 'windows of opportunity' in existing urban governance arrangements. The chapter concludes with a short summary of our findings and a discussion of a possible future agenda in this field of research.

Cross-Movement Alliances as a Novel Form of Agency in Urban Climate Governance

In the face of environmental degradation and increasing social inequalities new coalitions of social movements have emerged over the past three decades, converging around the notion of climate justice and calling for alternative visions of political organization, epitomized by the popular slogan 'system change not climate change' (see e.g. Patterson & Smith, 2017; Smith, 2014: 121). Combining the ecological agenda with a concern for widening social inequalities, global alliances such as the People's Climate Movement and the World Social Forum have become important channels for articulating contemporary critiques of neo-liberal globalization and unsustainable economic development. Especially across Europe and the United States, the rise of aggressive political populism and neo-liberal urbanization has contributed to the formation of novel political alliances between environmentalist, labour, indigenous, and decolonial movements; movements for gender equality; as well as between LGBTQI and other rights-based activist groups, for example in the context of the Right to the City Alliance (2007) or the Dakota Access Pipeline protests which began in early 2016.

While these developments show that environmental justice is tightly linked to other socially relevant issues such as political-economic critique or struggles against racially motivated violence, it must be noted that the building of alliances between different social movements is, in fact, not a new phenomenon. Historical examples for collaborative efforts between environmentalist and other social movements include the formation of the environmental justice movement as well as alliances between environmentalists and peace activists during the 1970s. Yet, there is evidence to suggest that cross-movement organizing reached a new quality during the 1990s, when the so-called 'Third Wave' of global environmentalism came to the fore. Alienated by an environmentalist agenda that was seen to be elitist in nature, co-opted by corporate interests and largely indifferent to the problems of the poor and marginalized, the environmental justice movement began to understand 'environment, race, and social justice issues as one complex' while aiming to form a diverse multi-issue environmental movement based on principles of inclusiveness and collaboration (Best, 2012: 68).

It is hardly surprising that many of these new cross-movement alliances were based in metropolitan centres and, at the same time, distinctly urban in their focus. The city served an important reference point for social movements and activists who were concerned about the widening gulf of inequality, racial instability, and poverty which is still characteristic of various urban spaces around the world. Supported by a large body of research that documented the struggles of indigenous communities, people of colour and the urban poor, 'Third Wave' environmentalists saw the city not only as a catalyst for political mobilization but also as a potent symbol of a finance-driven global economy seemingly unable to address the mounting pressures of social, racial, and environmental inequality (Behrens & Robert-Nicoud, 2014; McGranahan et al., 2016). Today, in the so-called 'urban century', these converging problems are certainly more pressing than ever. A recent study of racial differences in economic opportunity in the United States using longitudinal data covering nearly the entire US population from 1989 to 2015 shows, for example, that race remains a defining factor for upward social mobility, since 'black and white boys have very different outcomes even if they grow up in two-parent families with comparable incomes, education, and wealth, live on the same city block, and attend the same school' (Chetty & Hendren, 2018: 6; Chetty et al., 2018). Owing to their important role as political hubs and key centres for economic growth, cities are also conducive environments for private sector investment in public services. In many cases, however, privatization has led to poorer performance and working conditions, while also exacerbating social inequality (Rossi & Vanolo, 2015). From the glittering skyline of Manhattan's financial district to the inner-city slums of Johannesburg, the city thus embodies the contradictory tendencies of globalized urbanization and economic development.

Many of these critical views about the socioeconomic injustices of urbanization were also shared by the first advocates of a more environmentally focused 'urban' political ecology, a term that has been popularized by the Belgian geographer Eric Swyngedouw (1996) in his seminal article on the 'city as a hybrid'. By criticizing the underlying nature/culture dichotomy in urban research, and by arguing that the debate on the urban hitherto lacked a proper focus on the political dynamics through which the environment became urbanized, Swyngedouw set the tone for much of the subsequent work in UPE that has been published over the past two decades. Since then, urban political ecologists have investigated the politics of social and environmental (in)justice in cities (Heynen et al., 2018; Myers, 2008), engaged in attempts at decolonizing theory and practice (Lawhon et al., 2014; Leff, 2015; Schulz, 2017) and further developed the concept of the 'urban metabolism' to analyse politicized flows of resources in urban socio-technical systems (Gandy, 2004; Monstadt, 2009; Swyngedouw, 2006).

However, these critical accounts of urbanization notwithstanding, cities have typically been associated with ideas of progress, competitiveness, innovation, and creative renewal. Yet, there is growing consensus among policymakers, practitioners and researchers that the unprecedented successes in urban poverty reduction which have, for example, been documented in China over the past 40 years were achieved at the cost of strong social polarization and large-scale environmental deterioration (Guan et al., 2018). In the context of urban climate governance and sustainability transformations, cities around the world are now beginning to experiment with novel adaptation and mitigation strategies as part of their climate agendas, while social, economic, and cultural aspects of vulnerability reduction are increasingly moving into focus (Anguelovski & Carmin, 2011). As a result of these political developments, new actor alliances are emerging at a more formal level of urban climate governance, pointing to an ongoing cross-fertilization between more locally situated forms of political activism and more globally oriented constellations of actors. For example, recent years witnessed a growing interest, especially among governance scholars, in the agency of transnational city-networks and non-state actors (Acuto & Rayner, 2016; Chan et al., 2015; Lee, 2015; van der Ven et al., 2017). It has been stressed in this respect that the current concentration of transnational municipal networks in Europe and North America has been accompanied by considerable problems of political representation (Bansard et al., 2017) and that current orchestration efforts under the umbrella of the United Nations are still suffering from significant shortfalls in terms of democratic legitimacy and accountability (Bäckstrand & Kuyper, 2017). In addition, a comprehensive study of the 21 climate change coalitions which together form the environmental movement in the United States has shown that the overall network structure in this domain is thoroughly dominated by three large coalitions of non-state actors, namely 350.org, the U.S. Climate Action Network and TckTckTck (Schaefer Caniglia et al., 2015: 250–252). It is particularly interesting to note that the three coalitions who control practically all the resources within the US environmental movement also firmly embrace the paradigm of ecological modernization.

Against this background, it seems optimistic to assume that the aforementioned actor coalitions will reinvigorate urban climate governance within a timeframe that is sufficient to prevent the most detrimental consequences of climate change, or that they will be able to effectively address some of the well-known democratic deficits of top-down orchestration and technocratic entrepreneurialism in urban politics (for a summary, see MacLeod, 2011). Quite strikingly, the key problem that remains at the heart of urban (climate) politics has already been outlined by the American sociologist Harvey L. Molotch during the late 1970s. In his seminal paper titled "The City as a Growth Machine," Molotch (1976: 310, 325) aptly summarizes the dominant urban development paradigm by stating that the desire for industrial and economic growth 'provides the key operative motivation toward consensus for members of politically mobilized local elites, however split they might be on other issues', all while "the fear of unemployment acts to make workers politically passive (if not downright supportive) with respect to land-use policies, taxation programs, and antipollution nonenforcement schemes which, in effect, represent income transfers from the general public to various sectors of the elite." In addition, Hodson and Marvin (2010: 311) further contribute to the analysis of current trends in urban climate governance by asking whether ecological modernization, as 'the dominant logic of neoliberal responses is about the creation of "bounded" security in new ecological enclaves for premium users that ignore wider distributional questions about uneven access to resource politics (...) and build strategic protection from climate change and wider resource constraints'. Unfortunately, as it were, we may assume that climate change, variability, and environmental degradation will, at least in the near future, be quite 'democratic' in their negative effects.

As unsettling as these persistent trends in urban governance may be, they nevertheless lead us to the central purpose of this chapter, namely to identify the conditions under which newly emerging cross-movement alliances in urban climate governance have been able to exercise their political agency and achieve more socially and environmentally just outcomes. To begin with, it is evident that various social initiatives have been forged at the grassroots level of urban politics over the past two decades, often as a response to neo-liberal investment strategies such as the privatization of urban utility systems or the gentrification of urban and suburban spaces. Moreover, as adaptation and decarbonization strategies are becoming increasingly important factors to legitimize ecological modernization interventions in urban socio-technical systems, many local initiatives are now addressing environmental justice concerns in their endeavours to defend the public 'right to the city' (Harvey, 2012). A growing number of new international municipalist movements and civic platforms such as Barcelona en Comù, Refuge Cities and Fearless Cities are currently experimenting with democratic practices while attempting to build emancipatory political alternatives from the ground up. These neo-municipalist alliances are committed to increase access to common goods such as water, energy, healthcare, or housing, and oppose austerity politics, privatizations, and radical identity politics. Simultaneously, they also see themselves at the forefront of local actions against climate change. Whether such neo-municipalist networks and locally embedded cross-movement alliances against the privatization of public goods may be more successful in their efforts to promote social and climate justice than more established municipal networks such as ICLEI – Local Governments for

Sustainability (formerly named International Council for Local Environmental Initiatives) or the C40 Cities Climate Leadership Group is a question that still requires further research.

At present, it is nevertheless evident that the city as a living laboratory for societal innovation and creative transformations has caught the attention of UPE and governance scholars. Related emancipatory perspectives range from rather organized and policy-oriented 'urban experiments' (Bulkeley et al., 2015) to more 'radical' forms of democracy and urban politics from below (Loftus, 2012). Yet, our review of the existing literature in the fields of urban climate governance and UPE shows that scholars have diverted far less attention to the study of cross-movement alliances in urban climate governance than one may initially assume. Even if we turn our attention to the well-established field of social movement studies, it becomes clear that the topic of cross-movement alliances in urban climate governance has received relatively little attention in comparison to studies that either focus on cross-movement convergences between different urban social movements or deal with the importance of environmental movements for climate governance in general. Thus, with some notable exceptions (Bulkeley et al., 2014; Chatterton et al., 2013; Roberts & Parks, 2009), there is a need for more targeted research on cross-movement coalition building in urban climate governance that looks beyond highly formalized transnational municipal networks and focuses more clearly on movement convergence among social, environmental, and climate justice activism. In the following section, we aim to address this existing research gap by illuminating the specific circumstances under which cross-movement alliances in Accra and Berlin have been able to exercise their agency in the governance of urban socio-technical systems.

Negotiating the Governance of Urban Socio-Technical Systems: The Role of Cross-Movement Alliances in Accra and Berlin

Contemporary research clearly demonstrates that uneven spatial developments, precarity, and poverty in urban areas pose significant risks for already vulnerable populations and negatively affect the implementation of urban mitigation and adaptation strategies (United Nations, 2015; Whitehead, 2013). The exact causes for global trends such as increasing social and economic inequality have been much debated among scholars from various backgrounds and, at least in part, been attributed to the rise of a so-called 'neo-liberal' paradigm in urban policy and governance. Neo-liberalism, as an economic paradigm for urban development, is commonly associated with the ascent of conservative governments in the United States and the United Kingdom during the 1980s (Rossi & Vanolo, 2015). Considering the inherent historical complexities of so-called neo-liberal urbanization, Mayer (2017) points out that today's uneven geographies of the city are fundamentally shaped by the privatization and securitization of urban socio-technical systems, including public goods, services, and spaces. At the same time, historians point to a long and intricate history of persistent socio-economic inequality that far exceeds the effects of contemporary neo-liberal urbanization and warrants a more detailed and long-term analysis of human development trajectories (Scheidel, 2017).

Therefore, considering the multiple angles from which historically complex trajectories of urban development may be assessed, the cases of Berlin and Accra are certainly illustrative of the fact that studies dealing with the agency of cross-movement alliances in urban climate governance would risk painting a simplistic picture by focusing on a single dominant cause for citizen involvement, for instance resistance against neo-liberal urbanization. Instead, there is ample evidence to suggest that motivations behind cross-movement political engagement are not only highly diverse and context dependent, but also strategically articulated. Political goals such as defending the public 'right to the city' and resistance against privatization may indeed play an important role in some instances, as the Berlin case demonstrates. Yet, there are also examples in which the involvement of cross-movement alliances in the governance of urban socio-technical systems is characterized by entirely different political contexts and strategic motivations, for example in the case of Accra.

Negotiating the Governance of the Urban Energy Transition in Berlin

With a population of 3.7 million, Germany's capital, Berlin, is the largest metropolitan area in the country and the second most populous city in the European Union, covering an area of 891.7 square kilometres. Over the past 10 years, there has been ample evidence for increased cross-movement political engagement in the city as political struggles over the governance and ownership of basic infrastructures such as energy systems have become more intense. In view of pressing concerns about environmental degradation and climate change, these struggles can be interpreted as attempts to fundamentally alter the way in which energy is produced, distributed, and used. At the same time, it is important to note that recent political debates about the future of energy supply in Germany take place against the backdrop of the so-called 'Energy Transition' (Energiewende). Policies related to the Energiewende include the aim to phase out nuclear power in Germany by 2022, the immediate aim to reduce greenhouse gas emissions by at least 40 per cent in 2020 (relative to 1990 emission levels), as well as the aim to achieve a renewable energy share of 60 per cent in gross final energy consumption by 2050 (German Federal Ministry for Economic Affairs and Energy, 2016).

In this unique political climate, a citizen-led alliance emerged in 2011 with the aim to challenge the privatization of Berlin's urban energy system, and to reclaim public ownership and control. In order to better understand the aims and motives of this cross-movement alliance it is helpful to keep in mind that energy infrastructure is not only an agglomeration of physical artefacts such as power grids or energy production plants. Infrastructures, conceptualized as socio-technical systems, also encompass social actors that are using, maintaining, and regulating these infrastructures (Hummel & Kluge, 2004). Accordingly, a focus on infrastructure can be useful to theorize and understand urban transitions (Ernstson et al., 2010). However, until recently the governance and provision of infrastructures has primarily been perceived as a security, engineering, and administrative challenge, despite the significance of infrastructure systems for urban and regional sustainability (Hodson & Marvin, 2010; Monstadt, 2007). The dynamics of urban infrastructure and the challenges emerging from conflicts over the political economy of the city have not been fully explored (Bulkeley et al., 2010: 29).

In the case of Berlin, the protest against the privatization of the municipal energy system was spearheaded by the so-called Berlin Energy Roundtable (Berliner Energietisch), a cross-movement coalition of 56 different actors ranging from welfare and tenant organizations, environmental protection groups, church groups, and non-governmental organizations (NGOs) such as attac Berlin to smaller citizen-led initiatives. The main aim of this cross-movement alliance which started in 2011 was the re-municipalization or the 'reclaiming' of energy infrastructure 'from below' to support climate justice. In the words of the coalition:

> ... from energy access to climate justice and from anti-privatisation to workers' rights, people across the world are taking back power over the energy sector, kicking-back against the rule of the market and reimagining how energy might be produced, distributed and used. For many movements involved in struggles around energy, the concept of energy democracy is proving increasingly useful as a means of bringing together disparate but clearly linked causes under a shared discourse and, possibly, something of a common agenda.
>
> *(Berliner Energietisch, n.d.)*

As historical legacies and path dependencies play a significant role in any transition, the historic relevance of municipally owned utilities in Germany must be underlined. The German landscape of utilities has historically been dominated by strong public municipal companies which provided comprehensive services to the municipalities they served. Since the 1990s, however, a tendency towards privatization has become tangible. In Berlin, the former energy utility BEWAG was sold in 1997 to Vattenfall because of the tremendous debt of the city state. The privatization of BEWAG took place against the backdrop of a general paradigmatic shift towards economic liberalization and privatization in Germany,

Europe, and internationally, as critical urban infrastructure networks such as water and waste, energy, telecommunications, and much of the transport infrastructure were gradually opened up to private sector participation (Graham & Marvin, 2002). This process also led to a decrease of governance capacity at the city level in favour of multinational companies.

Yet, a shift towards the re-municipalization of urban socio-technical systems, especially of energy and water infrastructure, can be observed in recent years. Just as the paradigmatic shift towards privatization in the 1990s embodied the rationale of cutting costs and increasing economic efficiency through outsourcing, re-municipalization includes a variety of drivers and motivations. German municipalities, especially those with a weak economic performance, perceive decarbonization efforts as a chance for economic development, especially since feed-in tariffs incentivize investment in renewable energy technologies. Many municipalities are now actively pursuing decarbonization and climate mitigation strategies that have also been demanded by citizen coalitions.

The founding of the Berlin Energy Roundtable in 2011 is a primary example for such a citizen-led coalition motivated by political demands for the re-municipalization of privatized energy systems. One of the main political goals of this new cross-movement coalition is to challenge top-down economic and political power relations by demanding access to – and public control over – Berlin's energy grids in order to advance the transition towards a more democratic, sustainable and socially just energy system. In 2013, a unique window of opportunity for re-municipalization of Berlin's energy system opened up, when concession contracts for the city's energy and gas supply networks were about to expire. The expiring contracts for these services, along with public pressure to meet Berlin's self-declared target of becoming a carbon-neutral city by 2050 were both important factors that led to the formation of the alliance. Yet, the Energy Roundtable does not only focus on the re-municipalization of Berlin's energy systems. Its members also seek to advance democratic transparency, social accountability through responsible pricing policies, and improved citizen participation by proposing the establishment of a citizen-focused municipal power company (Bürgerstadtwerk) with a Board of Directors that is partly elected by the public (Becker et al., 2015: 299). The alliance also proposed that the Berlin municipal power company should rely on decentralized renewable energy plants in the Berlin–Brandenburg region to further Berlin's climate target of achieving carbon neutrality by 2050 (Berliner Energietisch, n.d.).

Overall, the key reasons for the ongoing re-municipalization trend in Germany can be summarized by pointing to the general increase in the costs of privatization, along with the dissatisfaction of citizens and their desire to have more options to manage and control their own utilities (Friedländer, 2013). In Berlin, these political demands led to a public referendum in 2013, with the main goal of re-municipalizing local utilities. Despite the fact that this key political goal was opposed by the city government, results demonstrated that 83 per cent of voters were in favour of re-municipalization efforts. However, the referendum was still not entirely successful, as the necessary turnout quorum of 25 per cent was missed by an extremely narrow margin. Critics attribute the low turnout rate to the city government's refusal to allow the referendum to be held on the same day as the federal elections (Becker et al., 2015: 299). Regardless of the pushback of the city government and other actors, the formation of novel cross-movement alliances like the Berlin Energy Roundtable is nevertheless indicative of a new form of agency in urban climate governance. Windows of opportunity such as the favourable political conditions created by Germany's energy transition and expiring concession contracts for the energy supply network enabled the Berlin Energy Roundtable to fundamentally challenge the top-down governance of the city's energy grid. Therefore, the Berlin Energy Roundtable alliance sent a clear political signal that could neither be ignored by Berlin's federal state government nor by referendum opponents.

Negotiating the Governance of Urban Socio-Technical Systems in Accra

Located at the Gulf of Guinea, Ghana's capital Accra is particularly vulnerable to climate hazards such as floods, landslides, coastal erosion and sea level rise which are likely to have severe impacts on critical infrastructure. With a current daytime population of approximately 3.5 million, the growing Accra metropolitan area also faces considerable challenges in terms of sustainable and inclusive urban development in sectors such as water, sanitation, energy, and housing (Owusu & Oteng-Ababio, 2015: 10). Overall, climate change poses a formidable risk for Ghana's predominantly agrarian economy and may jeopardize food security as well as development gains in climate-sensitive sectors such as cocoa production, tourism, forestry, and hydro-electric power generation (NCCP, 2013). Together with persistent socioeconomic challenges, climate change threatens to increase health risks, place additional pressures on institutions and resources, and propel internal migration to urban centres. Furthermore, it is to be expected that additional expenditures for disrupted value chains, emergency relief, and post-disaster rehabilitation will significantly reduce the potential for economic growth as well as available funds for sustained poverty reduction and environmental rehabilitation. At the highest level of the Ghanaian government, climate change is therefore considered a serious problem caused by industrialized countries but also seen as a political 'window of opportunity' to address well-known development priorities.

Political debates about climate governance in Ghana are thus closely linked to familiar sustainable development discourses, although new financial instruments such as the Green Climate Fund are becoming increasingly influential with regard to the strategic orientation of urban development planning and disaster risk reduction. Simultaneously, discussions about the linkages between climate change and development are emerging in the news media, civil society, and the academe, while state institutions and international donor organizations are positioning themselves in the changing political landscape. Newly established knowledge networks such as the 'SDG13 CSO Platform' and the 'Kasa Initiative Ghana' also contribute to the public agenda-setting process by acting as knowledge brokers between smaller and more locally oriented civil society organizations and the state administration.

In comparison to the Berlin case, available research nevertheless suggests that new cross-movement alliances in Ghana emerged primarily as a result of top-down policy processes at the national level of climate governance (Schulz, 2017). Policy decisions which have a direct impact on urban climate governance and the development of socio-technical systems in Accra are, therefore, closely linked to the agenda-setting process at the national level of policy making. For example, participatory research during civil society stakeholder meetings prior to the official release of Ghana's National Climate Change Policy has shown that highly networked individuals in academia and NGOs such as Friends of the Earth and CARE International exerted significant political influence in the climate change policy arena (Schulz, 2017). While these findings may not be entirely surprising, they do not necessarily imply a democratic deficit either, as several thousand people were involved in the consultative process that led to the adoption of Ghana's National Climate Change Policy. These actors included not only the Ghanaian parliament, ministerial staff, and public sector employees, but also representatives of customary authorities, the media, civil society and community-based organizations, gender rights groups, the private sector, international and faith-based organizations, as well as academic institutions and the donor community. On the one hand, this complex consultation process under the guidance of the state administration supported the formation of new cross-movement alliances between various civil society groups and resulted in a progressive and socially balanced policy document. On the other hand, the process also revealed the difficulties of meaningful participation in view of limited institutional capacities and resources. In other words, the fact that civil society actors are *in principle* able to exercise their agency and influence climate policy does not mean that all of them have the same capacity to do so.

It follows from this example that the challenges for civil society participation which arise as a result of institutional limitations cannot be easily understood without a thorough analysis of the systemic and contextual factors of urban climate governance in Accra. As long as policy outputs do not necessarily translate into policy outcomes, there is ample reason to assume that the combined agency of newly emerging cross-movement alliances is primarily symbolic. Similar findings have also been reported by Morinville and Harris (2014: 1), who investigated participatory mechanisms in the context of urban water governance in Accra and emphasize that 'discerning the potential and limits for sustainable resource governance and associated development goals requires that participatory mechanisms be subjected to systematic and contextual analysis'. Thus, while independent cross-movement alliances in Berlin were highly successful in exploiting political windows of opportunity to influence political outcomes in opposition to the local government, the Ghanaian case shows that even the most egalitarian and participatory collaboration between civil society and government actors is potentially ineffective as long as resources are lacking, and policy implementation remains uncertain.

Conclusion

In sum, our findings indicate that research on the agency of cross-movement alliances in urban climate governance would risk painting a simplistic picture by focusing on a single dominant cause for citizen involvement such as resistance against neo-liberal urbanization. Instead, there is strong evidence to suggest that motivations behind cross-movement political engagement are not only diverse and context dependent, but also strategically articulated. Nevertheless, the question whether neo-municipalist networks and citizen-led alliances are more successful in advancing social and climate justice than established municipal networks remains open. There is certainly a need for more focused research on cross-movement coalition building in urban climate governance that looks beyond formalized transnational municipal networks and investigates cross-movement convergence among social, environmental, and climate justice activism. Broadly speaking, the formation of new cross-movement alliances in urban climate governance points to an ongoing cross-fertilization between more locally situated forms of urban political activism and more globally oriented constellations of actors. Moreover, concrete examples from Accra and Berlin show that neither top-down and state-centric, nor bottom-up and citizen-led forms of organizing are necessarily more effective in terms of achieving desired political outcomes, regardless of who defines these outcomes.

Current discourses about urbanization and climate governance in Accra, for example, primarily revolve around development-related aspects such as basic service provision or coverage and are often supported by arguments in favour of privatization, mainly because public service provision is regarded as insufficient to meet the needs of a growing population. Research in the context of the public consultation that took place prior to the release of Ghana's National Climate Change Policy also revealed that highly networked individuals with strong political ties to the state administration and international donor organizations exert a considerable amount of influence on decision making. Yet, while smaller community-based organizations must often rely on the expertise or technical know-how of influential individuals and larger NGOs to make their voices heard, the state-led consultation process also formalized the exchange between various groups and led to the emergence of new actor coalitions. Thus, owing to the growing diversity of individuals and organizations who are involved in Ghanaian climate governance, it becomes increasingly important to study cross-sectoral policy networks at the national level that are having a notable effect on the governance of urban socio-technical systems and are spearheading policy implementation across scales. One of the key challenges with regard to the practical side of top-down climate governance in Accra is certainly the creation of a favourable political environment for the self-empowerment of marginalized groups and the inclusive implementation of climate policies that simultaneously contribute to poverty reduction.

In comparison to the case of Accra, the situation in Berlin is rather indicative of how cross-movement organization and political agency may also emerge from the bottom up. The formation of a citizen alliance with the clear goal to develop and implement emancipatory alternatives to neo-liberal urbanization demonstrates that reclaiming the 'right to the city' is not merely an academic idea. It can also serve as a concrete motivation for cross-movement alliances to become a publicly visible counterforce against the top-down governance of socio-technical systems by powerful economic and political actors. The political critique that the city government of Berlin actively prevented the public referendum from reaching the necessary quorum by not allowing it to be held on the same day as the federal elections does not necessarily mean, however, that re-municipalization efforts in Germany are to no avail. Similar alliances have also formed in other German cities such as Hamburg, where citizens successfully resisted the ongoing privatization of public goods and services. In addition, the Berlin Energy Roundtable remains active in international campaigns for climate justice and cooperates with other alliances across Europe to advocate for climate justice and energy democracy.

In view of these findings, we therefore suggest that future research on the agency of cross-movement alliances needs to focus more closely on contextual factors such as historical path dependencies, political gridlocks, available resources and time constraints, as these factors determine how effectively actors may use or create political 'windows of opportunity' to advance their normative goals. The use of the word effectiveness, in this context, does not necessarily connote a managerial or Eurocentric understanding of politics.

It rather indicates that there is no guarantee for achieving normatively desired outcomes (i.e. 'effective' action), even if political windows of opportunity are opening up. Ultimately, we propose that the frames of reference used by researchers to define what constitutes a 'movement' and analyse political processes should be continuously subjected to critical scrutiny, especially to avoid the imposition of overly rigid, Eurocentric or externally devised categories on actor constellations that are characterized by a high degree of plurality, diversity, and flexibility.

References

Acuto, M., & Rayner, S. (2016). City networks: Breaking gridlocks or forging (new) lock-ins? *International Affairs*, 92(5): 1147–1166.

Amoako, C. (2017). Emerging grassroots resilience and flood responses in informal settlements in Accra, Ghana. *GeoJournal*, 1–17. DOI:10.1007/s10708-017-9807-6.

Anguelovski, I., & Carmin, J. (2011). Something borrowed, everything new: Innovation and institutionalization in urban climate governance. *Current Opinion in Environmental Sustainability*, 3(3): 169–175.

Bäckstrand, K., & Kuyper, J. (2017). The democratic legitimacy of orchestration: The UNFCCC, non-state actors, and transnational climate governance. *Environmental Politics*, 26(4): 764–788.

Bansard, J. S., Pattberg, P. H., & Widerberg, O. (2017). Cities to the rescue? Assessing the performance of transnational municipal networks in global climate governance. *International Environmental Agreements: Politics, Law and Economics*, 17(2): 229–246.

Becker, S., Beveridge, R., & Naumann, M. (2015). Remunicipalization in German cities: Contesting neo-liberalism and reimagining urban governance? *Space and Polity*, 19 (1): 76–90.

Becker, S., Blanchet, T., & Kunze, C. (2016). Social movements and urban energy policy: Assessing contexts, agency and outcomes of remunicipalisation processes in Hamburg and Berlin. *Utilities Policy*, 41: 228–236.

Behrens, K., & Robert-Nicoud, F. (2014). Survival of the fittest in cities: Urbanisation and inequality. *The Economic Journal*, 124(581): 1371–1400.

Berliner Energietisch (n.d.). *English Information*. http://berliner-energietisch.net/english-information (accessed 5 April 2018).

Best, S. (2012). Greening philosophy. In S. Fassbinder, A. Nocella, & R. Kahn (eds.), *Greening the Academy: Ecopedagogy through the Liberal Arts*. Rotterdam: Sense Publishers, 63–76.

Bulkeley, H., Castán Broto, V., & Edwards, G. (2015). *An Urban Politics of Climate Change: Experimentation and the Governing of Socio-Technical Transitions*. Abingdon: Routledge.

Bulkeley, H., Edwards, G. A., & Fuller, S. (2014). Contesting climate justice in the city: Examining politics and practice in urban climate change experiments. *Global Environmental Change*, 25: 31–40.

Bulkeley, H., Castán Broto, V., & Maassen, A. (2010). Governing urban low carbon transitions. In H. Bulkeley, V. Castán Broto, M. Hodson, & S. Marvin (eds.), *Cities and Low Carbon Transitions*. London: Routledge, 29–41.

Carroll, W. K., & Ratner, R. S. (1996). Master framing and cross-movement networking in contemporary social movements. *The Sociological Quarterly*, 37(4): 601–625.

Castán Broto, V., & Bulkeley, H. (2013). Maintaining climate change experiments: Urban political ecology and the everyday reconfiguration of urban infrastructure. *International Journal of Urban and Regional Research*, 37(6): 1934–1948.

Chan, S., van Asselt, H., Hale, T., et al. (2015). Reinvigorating international climate policy: A comprehensive framework for effective nonstate action. *Global Policy*, 6(4): 466–473.

Chatterton, P., Featherstone, D., & Routledge, P. (2013). Articulating climate justice in Copenhagen: Antagonism, the commons, and solidarity. *Antipode*, 45(3): 602–620.

Chetty, R., & Hendren, N. (2018). Race and economic opportunity in the United States: executive summary. *The Equality of Opportunity Project (EOP)*. Retrieved 5 April 2018, www.equality-of-opportunity.org/assets/documents/race_summary.pdf (accessed 5 April 2018).

Chetty, R., Hendren, N., Jones, M. R., & Porter, S. R. (2018). Race and economic opportunity in the United States: An intergenerational perspective (No. w24441). Cambridge, MA: National Bureau of Economic Research.

Ernstson, H., Leeuw, S. E. V. D., Redman, C. L., et al. (2010). Urban transitions: On urban resilience and human-dominated ecosystems. *AMBIO: A Journal of the Human Environment*, 39(8): 531–545.

Friedländer, B. (2013). Rekommunalisierung öffentlicher Dienstleitungen: Konzept–Entwicklungstendenzen–Perspektive. Arbeitspapier No. 45, Institut für öffentliche Finanzen und Public Management, Finanzwissenschaft. Universität Leipzig.

Gandy, M. (2004). Rethinking urban metabolism: Water, space, and the modern city. *City*, 8 (3): 363–379.

German Federal Ministry for Economic Affairs and Energy. (2016). Fifth "energy transition" monitoring report. The energy of the future (2015 reporting year). Berlin. www.bmwi.de/Redaktion/EN/Publikationen/fuenfter-monitoring-bericht-energie-der-zukunft.pdf?__blob=publicationFile&v=2 (accessed 5 April 2018).

Graham, S., & Marvin, S. (2002). *Splintering Urbanism: Networked Infrastructures, Technological Mobilities and the Urban Condition*. London: Routledge.

Guan, X., Wei, H., Lu, S., Dai, Q., & Su, H. (2018). Assessment on the urbanization strategy in China: Achievements, challenges and reflections. *Habitat International*, 71: 97–109.

Harvey, D. (2012). *Rebel Cities: From the Right to the City to the Urban Revolution*. London: Verso Books.

Heynen, N. (2014). Urban political ecology I: The urban century. *Progress in Human Geography*, 38(4): 598–604.

Heynen, N., Aiello, D., Keegan, C., & Luke, N. (2018). The enduring struggle for social justice and the city. *Annals of the American Association of Geographers*, 108(2): 301–316.

Hodson, M., & Marvin, S. (2010). Urbanism in the Anthropocene: Ecological urbanism or premium ecological enclaves? *City*, 14(3): 298–313.

Hummel, D., & Kluge, T. (2004). Das Konzept Gesellschaftliche Naturverhältnisse. In BMBF (ed.), *Steuerung und Transformation. Überblick über theoretische Konzepte in den Projekten der sozial-ökologischen Forschung* (Diskussionspapier 1), 93–100.

Lawhon, M., Ernstson, H., & Silver, J. (2014). Provincializing urban political ecology: Towards a situated UPE through African urbanism. *Antipode*, 46(2): 497–516.

Lee, T. (2015). *Global Cities and Climate Change: The Translocal Relations of Environmental Governance*. London: Routledge.

Leff, E. (2015). The power-full distribution of knowledge in political ecology: A view from the South. In T. Perreault, G. Bridge, & J. McCarthy (eds.), *The Routledge Handbook of Political Ecology*. Abingdon: Routledge, 64–75.

Loftus, A. (2012). *Everyday Environmentalism: Creating an Urban Political Ecology*. Minneapolis: University of Minnesota Press.

MacLeod, G. (2011). Urban politics reconsidered: Growth machine to post-democratic city? *Urban Studies*, 48(12): 2629–2660.

Mayer, M. (2017). Whose city? From Ray Pahl's critique of the Keynesian city to the contestations around neoliberal urbanism. *The Sociological Review*, 65(2): 168–183.

McGranahan, G., Schensul, D., & Singh, G. (2016). Inclusive urbanization: Can the 2030 agenda be delivered without it? *Environment and Urbanization*, 28(1): 13–34.

Molotch, H. (1976). The city as a growth machine: Toward a political economy of place. *American Journal of Sociology*, 82(2): 309–332.

Monstadt, J. (2007). Urban governance and the transition of energy systems: institutional change and shifting energy and climate policies in Berlin. *International Journal of Urban and Regional Research*, 31(2): 326–343.

Monstadt, J. (2009). Conceptualizing the political ecology of urban infrastructures: Insights from technology and urban studies. *Environment and Planning A*, 41(8): 1924–1942.

Morinville, C., & Harris, L. (2014). Participation, politics, and panaceas: Exploring the possibilities and limits of participatory urban water governance in Accra, Ghana. *Ecology and Society*, 19(3): 36.

Myers, G. A. (2008). Sustainable development and environmental justice in African cities. *Geography Compass*, 2(3): 695–708.

NCCP. (2013). National Climate Change Policy. Accra: Ministry of Environment, Science, Technology and Innovation (MESTI). http://www.un-page.org/files/public/ghanaclimatechangepolicy.pdf (accessed 5 April 2018).

Owusu, G., & Oteng-Ababio, M. (2015). Moving unruly contemporary urbanism toward sustainable urban development in Ghana by 2030. *American Behavioral Scientist*, 59 (3): 311–327.

Patterson, J., & Smith, J. (2017). Environmental justice initiatives for community resilience. In B. Schaefer Caniglia, M. Vallée, & B. Frank (eds.), *Resilience, Environmental Justice and the City*. Abingdon: Routledge, 216–233.

Rice, J. L. (2014). An urban political ecology of climate change governance. *Geography Compass*, 8(6): 381–394.

Roberts, J. T., & Parks, B. C. (2009). Ecologically unequal exchange, ecological debt, and climate justice: The history and implications of three related ideas for a new social movement. *International Journal of Comparative Sociology*, 50(3–4): 385–409.

Rossi, U., & Vanolo, A. (2015). Urban neoliberalism. In J. D. Wright (ed.), *International Encyclopedia of the Social & Behavioral Sciences*, 2nd edn. Amsterdam: Elsevier, 846–853.

Schaefer Caniglia, B., Brulle, R. J., & Szasz, A. (2015). Civil society, social movements, and climate change. In R. E. Dunlap, & R. J. Brulle (eds.), *Climate Change and Society: Sociological Perspectives*. Oxford: Oxford University Press, 235–268.

Scheidel, W. (2017). *The Great Leveler: Violence and the History of Inequality from the Stone Age to the Twenty-First Century*. Princeton, NJ: Princeton University Press.

Schulz, K. A. (2017). Decolonizing political ecology: Ontology, technology and 'critical' enchantment. *Journal of Political Ecology*, 24(1): 125–143.

Silver, J. (2014). Incremental infrastructures: Material improvisation and social collaboration across post-colonial Accra. *Urban Geography*, 35(6): 788–804.

Simpson, C. R. (2015). Multiplexity and strategic alliances: The relational embeddedness of coalitions in social movement organisational fields. *Social Networks*, 42: 42–59.

Smith, J. (2014). Counter-hegemonic networks and the transformation of global climate politics: Rethinking movement-state relations. *Global Discourse*, 4(2–3): 120–138.

Swyngedouw, E. (1996). The city as a hybrid: On nature, society and cyborg urbanization. *Capitalism Nature Socialism*, 7(2): 65–80.

Swyngedouw, E. (2006). Metabolic urbanization: The making of cyborg cities. In N. Heynen, M. Kaika, & E. Swyngedouw (eds.), *In the Nature of Cities: Urban Political Ecology and the Politics of Urban Metabolism*. London: Routledge, 21–40.

United Nations. (2016). Habitat III: New urban agenda. Quito. http://habitat3.org/wp-content/uploads/Habitat-III-New-Urban-Agenda-10-September-2016.pdf (accessed 5 April 2018).

United Nations. (2015). *UN Habitat Climate Change Strategy (2014–2019)*. Nairobi. https://unhabitat.org/un-habitat-climate-change-strategy-2014–2019/# (accessed 5 April 2018).

van der Ven, H., Bernstein, S., & Hoffmann, M. (2017). Valuing the contributions of nonstate and subnational actors to climate governance. *Global Environmental Politics*, 17(1): 1–20.

Whitehead, M. (2013). Neoliberal urban environmentalism and the adaptive city: Towards a critical urban theory and climate change. *Urban Studies*, 50(7): 1348–1367.

7

The Politics of Data-Driven Urban Climate Change Mitigation

SARA HUGHES, LAURA TOZER, AND SARAH GIEST

Introduction

The calls for data-driven decision-making in cities are growing, to the point that many commentators and stakeholders now expect cities to use data to govern. This trend encompasses several buzzwords, such as 'big data', which refers to data that is high volume, high velocity, and high in variety (Batty, 2016; Laney, 2001); and 'smart cities' or 'smart urbanism', which describes cities that are becoming more automated by using data generated in real time from sensors embedded in the environment (Batty, 2016; Marvin, Luque-Ayala, & McFarlane, 2016). Urban climate change mitigation policy has not been immune to the trend toward data-driven decision-making. From building energy use benchmarking programs to greenhouse gas (GHG) emissions inventories and vulnerability assessments, data play an important part in urban climate change policymaking. Accounting for GHG emission production, flow, and change is a central task in data-driven urban climate change mitigation. Using carbon calculation tools, the GHG emissions associated with a city and its citizens are frequently quantified, categorized, and interpreted to suggest appropriate response actions (Bulkeley, 2013).

City governments are no strangers to working with data, though this has been mostly traditional datasets such as resident surveys and administrative data collected internally. Over time, these datasets have begun to be complemented with more dynamic information, such as real-time tracking of traffic or energy consumption data that has expanded the scope and volume of data available to local decision-makers. The assumption that data can be the (or even *a*) primary input for decisions-making does not always reflect current challenges facing urban decision-makers. As in any policymaking process, cities rely on a variety of inputs to formulate and decide on climate mitigation policies. In addition to data (including expert reports and sometimes scientific committees) decision-makers are keenly aware of election cycles and the availability of financial resources. In early Earth System Governance (ESG) research a lack of data has typically been viewed as a barrier to action, and generating and incorporating data is often recommended as something cities should strive for. In this sense, data-driven decision-making is seen as a means of enhancing the agency and capacity of cities to act to address climate change, reflecting Michael Bloomberg's famous motto, "If you can't measure it you can't manage it" (e.g. Bloomberg Finance, 2018).

However, the shift to data-driven decision-making in urban climate governance intersects with and shapes the diverse agents of change in important ways. Indeed, scholars have begun to grapple more explicitly with the political and critical dimensions of data-driven decision-making in the urban context, in part by engaging with other communities such as science and technology studies and smart cities scholarship.

In this chapter we examine and synthesize the literature around data-driven decision-making for urban climate change mitigation, focusing specifically on the political dimensions of this trend. We push back against any remaining notions that data are a 'neutral' resource for decision-makers, that data remove political considerations from the policy process, and that they are relatively easy to access and handle. Research has refuted these claims, with Kitchin et al. (2017: 10) demonstrating that 'urban data are always cooked and never raw', implying that data are in fact not an apolitical entity that can be tapped into by city governments. In addition, when looking at the data available to policymakers, it becomes clear that extracting usable information from data for specific policy domains is quite complex and challenging. Data are often scattered throughout departments and largely incomplete owing to legal, financial, or technical hurdles.

We focus on three main strains of scholarship: the promotion of data as a policy and governance input, the politics of data as a mechanism for transparency and accountability, and the democratic implications of data-driven climate change mitigation through the empowerment and disempowerment of different agents in urban governance. Throughout the chapter we use examples from cities already working to incorporate new data sources into their climate change mitigation policy processes to provide further context to the concepts and issues highlighted in the literature. Our review reveals a strong critical perspective in the literature – led in many ways by ESG scholars – that should continue in combination with identifying and exploring productive ways forward for cities as data availability continues to grow. In addition, we advocate for research that delves more deeply into understanding the actors implicated in data collection and storage, and that leverages comparative analysis of cities undertaking data-driven climate change mitigation in different political-economic contexts.

Data as a Policy and Governance Input

Research has framed data as a technical input to decision-making, one which presents challenges in the form of the details of data analytics and the technical infrastructure available, and that ultimately enhances the *capacity* of city governments to develop and implement effective and efficient climate change policies. Some recent work has focused on the role such data inputs play in shaping the 'smart city' or 'smart governance' by 'crafting new forms of human collaboration . . . to obtain better outcomes and more open governance processes' (Meijer & Bolivar, 2016: 392). In the context of urban carbon governance and smart energy systems, 'low carbon transitions are increasingly being configured in relation to the notion of smart energy systems, while smart as a discourse, practice and set of interventions is also being related to the task of reconfiguring and transforming (urban) infrastructure' (Bulkeley et al., 2016: 1710; see also Vanolo, 2014). In this section we explore these themes in two parts: the emphasis on data for capacity building (a primary focus of the ESG literature) and the role of data in supporting 'smart governance' to combat the wicked problem of climate change (Levin et al., 2012).

Data and Capacity Building

An important measure of decision-making capacity is the ability to draw on relevant knowledge in the policy-making process (Howlett, 2009; Nutley et al., 2007; Pawson, 2006; Romero-Lankao, 2012; Sanderson, 2006). Capacity can be defined as the 'autonomy, resources and decision-making power of local authorities in relation to critical sources of GHG emissions and policy sectors', such as infrastructure or waste (Romero-Lankao, 2012: 20). Another take on capacity is that of 'smart governance capacity' (Edelenbos et al., 2017; Innes & Booher, 2003), which combines human, relational, and organizational capacity. Each capacity is linked to specific activities, such as boundary spanning at the individual level, developing trustworthy relationships, and coordinating and communicating at the organizational level. For example, the efforts of the World Council on City Data to standardize data collection efforts in cities is meant to empower decision-making and facilitate learning between cities (www.dataforcities.org/).

Capacity can apply to the capacity of civil servants to handle climate change data. This means they, for example, need a data science background or the support of IT personnel to make use of the data that are available. Further, there is the understanding that governmental departments require the capacity to store the data and share it based on interdepartmental agreements and privacy regulations. This implies that local departments know which data they collect, how they are stored (e.g. which format), and how and when they are allowed to share them with other (government) entities. Some cities decide on a one-off sharing model, where data are shared while a specific project is ongoing, but not before or after. This limits the time that shared data are available. Owing to privacy regulations, some cities also decide to collect data from scratch together with, for example, energy companies, in order to gain full access to consumption data (Giest, 2017). Finally, capacity applies to the technical capacity to store large amounts of data.

In these scenarios, government has the supply-side data, such as the location of roads or public transport schedules, and demand-side data are generated via cellphones or GPS trackers in real time. Combining the two to deliver public services in cities is, however, a challenge. As Tomer and Shivaram (2017) found, based on interviews with US public servants, 'public agencies simply do not have enough data scientists on staff or senior management experience to navigate a complete transition to big data platforms' (p. 9). Local governments lack data standards and data exchange policies as well as the hardware to deal with these data. In short, they lack the technical and skill capacity to deal with new types of data. Finally, this conceptualization of capacity is developed on the premise that there is a vast amount of data available to cities. This might not always be the case, especially if urban areas are smaller and they rely on the regional or national government to collect some of the environmental data (Giest, 2017).

To summarize, local capacity is seen to be critical to developing an urban response to climate change (Anguelovsky & Carmin, 2011; Romero-Lankao, 2012). For example, local governments have the ability to guide carbon emission reduction, because of the authority over land use, buildings, etc., but they are dependent on financial resources from national level. In addition, when governments experience low levels of data skills they risk incorporating scientific knowledge ineffectively into the decision-making process (Howlett, 2009). For urban climate information in particular, sophisticated climate change models give additional input for mitigation, but the models also add information that might expose decision-makers to irrelevant material and lead to prolonged administrative processes (Höchtl et al., 2016; Larson et al., 2015; Trenberth, 2010). The focus on gathering and archiving data has, in turn, limiting effects on building capacity. Costly data management linked to carbon emission reduction can thus pose a bottleneck for implementing and integrating climate change initiatives (French et al., 2015).

Capacity building was a primary motivation for New York City's energy use benchmarking legislation introduced in 2009. Greater awareness and detailed knowledge of energy use in the city's largest buildings was thought to be an effective means of building the capacity of, on the one hand, building owners to make smarter decisions about their buildings and operations and, on the other hand, of the city government to develop programs able to best target high-energy users in the city. At least 80 per cent of New York City's GHG emissions can be attributed to energy used in buildings; when the city realized this it also quickly became apparent that there were little if any data available on the patterns and distribution of that energy use. Requiring that building owners submit energy and water use data was a means for enhancing the capacity of both sets of actors. For example, the Mayor's Office of Sustainability writes in its description of energy and water benchmarking that 'with better analytical tools, the City aims to help property owners, buyers, and tenants to factor energy, water, and greenhouse gas emissions into their real estate decision making processes'.[1] However, significant investments of both human and financial resources have been required to support building owners in collecting and reporting these data and in establishing the capacity within city government to grapple with growing datasets.

Data and Smart Governance

Various scholars describe climate change as a so-called 'wicked' problem (Head & Halford, 2015; Rittel & Weber, 1973). Wicked is the notion that adverse social and environmental situations 'overwhelm existing practices and persist even after the application of best-known practice' (FitzGibbon & Mensah, 2012: 1). This describes a complex context in which there are dynamic and evolving issues and stakeholders do not necessarily agree on what the problem is or how to resolve it (Rittel & Weber, 1973). Some commentators and scholars have argued that the use of data would be a way to overcome some of the challenges linked to the complexities of urban governance and climate change.

The smart city movement in particular identifies the incorporation of technology and data as a way to identify and solve more problems effectively. Decision-makers have the potential to be less reactive and more proactive through the use of large datasets and predictive analytics (Mattern, 2013). Data can provide new perspectives on the temporal and spatial dimensions of urban systems and processes (Batty, 2016). The smart city concept treats cities as a set of manageable systems that behave in rational, mechanical, linear, and hierarchical ways and can, based on this, be steered and controlled (Kitchin, 2017). The 'smart governance' perspective, one element of the larger smart city approach, further looks at the inner workings of city governments. It takes on a citizen or user-centred perspective and highlights the connections made within governments to provide certain services (Kourtit et al., 2012). "The idea of collaborations is more central to this approach and authors

[1] https://serv.cusp.nyu.edu/projects/evt/ (accessed 15 November 2018).

focus on developing productive interactions between networks of urban actors" (Meijer & Bolivar, 2016: 397).

City dashboards are one example of the move towards smart governance. These applications allow local governments to display various types of data for the work of civil servants in one location. They include administrative data as well as statistical data from national institutions. In addition, such dashboards contain scientific data on environmental conditions, such as pollution or noise, crowd-sourced data by citizens, and social media data (Kitchin & McArdle, 2017). While these applications provide an overview of the status of city operations in a centralized way, they also act as translators for the original data. Policymakers acting on this information therefore run the risk of deciding based on ideologically framed and political information (Kitchin et al., 2016; Kitchin & McArdle, 2017).

In addition, each policy domain, such as environment, energy, water, or waste is itself complex and often governed in silos. This makes it difficult to actually implement a centralized dashboard.

The simplified notion of urban governance via smart city initiatives has been heavily criticized by several scholars who point towards the political and social dimensions of urban issues (Morozoy, 2013) and highlight that smart city initiatives treat cities alike and thus make such measures ahistorical and aspatial (Greenfield, 2013). The smart city perspective also overlooks some of the potentially profound social, political, and ethical consequences of implementing certain data measures linked to surveillance and privacy (Graham, 2005; Kitchin, 2014, 2017).

Another critique of the smart city research is that local governments are part of a multilevel system in which their decision-making power and access to data might be constrained by higher levels of authority at national or transnational level (Edelenbos et al., 2017).

In short, the complexity that many of the smart initiatives want to escape from by focusing on data- and technology-driven solutions further complicate urban governance owing to potentially profound consequences or the demand for additional stakeholders or knowledge. This is what Levin et al. (2012) termed a 'super wicked problem', where not only the urban issues are complicated and multifaceted, but also the solutions are unstandardized and highly experimental (Edelenbos et al., 2017). Linking this back to capacity, these complexities require cities to focus on setting up partnerships with stakeholders and integrating data infrastructures, which results in paying limited attention to the internal capacity to utilize data within government (Giest, 2017).

To summarize, recent research has become more critical of big data and the smart city and smart governance movement. The literature also increasingly focuses on the political and social dimensions of these new technologies. Two common themes that emerge from more recent publications are the capacity of government to utilize big data for urban measures and the problem of addressing a wicked problem with a complex solution. Data-driven urbanism provides a solution for some urban problems, but only within certain limitations (Kitchin et al., 2017). Tackling climate change requires the combination of data sources and the collaboration of various government departments. The danger in the data-driven urban governance perspective is thereby twofold: First, the breakdown of the city into technical and infrastructural elements is seen as a full picture of urban life (de Waal, 2017), and, second, that cities make decisions based on data that is not fully vetted for their potential social and political consequences.

Data as a Mechanism for Transparency and Accountability

In addition to emphasizing the role of data in building capacity and smart governance, research has highlighted the ability of data to serve as a mechanism for enhancing transparency and accountability in local decision-making processes and service provision (Goldsmith & Crawford, 2014). As in the case of capacity and opportunity building, however, using data towards such ends is not a politically neutral undertaking and brings the potential to marginalize those issues and stakeholders not captured by relevant datasets. Foregrounding transparency and accountability shifts the conversation from decision-making to broader processes and patterns of urban governance and the relationship among city governments, external stakeholders, the private sector, and urban residents.

Data-driven decision-making for climate change mitigation has the potential to improve the transparency and accountability of city governments and provide concrete metrics by which voters and stakeholders can evaluate services and, therefore, their satisfaction with municipal leaders and programs. Publicly available data and accounting methods can empower the public and other watchdogs to hold governments accountable for their emissions reduction targets and other climate change goals. Programs like energy use benchmarking, adopted by more than a dozen US cities, help to highlight major energy users in the city, providing an additional pathway for accountability (Cox, Brown, & Sun, 2013). However, there are several ways in which mobilizing data as a tool for enhancing accountability and transparency is an inherently political undertaking.

Politics of Accountability and Transparency

First, the metrics used for accountability often have more to do with a city's interest in maintaining legitimacy and authority than with environmental outcomes (Gordon, 2016; Kramarz & Park, 2016, 2017). Kramarz and Park (2016) point out that both the design of accountability institutions and the execution of interventions are important – otherwise, 'authority holders can be held to account for their actions without necessarily mitigating negative environmental impacts'. However, the metrics by which decision-makers are held accountable – whether reductions in GHG emissions, number of new programs initiated, or kilowatt hours of electricity saved – reflect particular logics of accountability and ultimately shape the conduct of cities. Kramarz and Park (2016) distinguish between public logics, defined as an interest in maintaining voter support and re-election; private logics, the interest in maintaining the confidence and support of the private sector and other sources of investment; and voluntary logics, or a kind of moral response to social rules and expectations for city government. Gordon (2016) also distinguishes between external and internal logics of accountability, in which cities may seek to be accountable to higher levels of government, transnational organizations, private investors, or financial institutions (external) or internally to citizens or the networks they participate in. Furthermore, the use of carbon and GHG emissions measurement and reporting tools by cities and other actors has been steadily increasing despite the relative lack of policy action. Knox-Hayes and Levy (2011) identified three core drivers of this trend: regulatory compliance, pressure from non-governmental organizations (NGOs) and participation requirements for carbon markets, and reducing energy costs and managing reputational risks. Embedded within these are two distinct institutional logics of carbon disclosure: a corporate logic of carbon risk management and carbon trading, and an NGO-oriented logic based on transparency and accountability. The authors argue that reporting and disclosure systems are headed towards a more corporate logic, which enhances the diffusion of disclosure but weakens it as a tool for driving substantial cuts in GHG emissions. Disclosure to various authoritative political, private, or voluntary institutions does not necessarily drive environmental action and, instead, 'being truly environmentally accountable is not only to be held responsible and answerable for one's actions, but to be responsible and answerable for protecting the environment and principal stakeholders' (Kramarz & Park, 2016). As a result, the content and standards of reporting tools, such as GHG emissions inventories, can become highly politicized and have historically been difficult to compare between cities. The increasing use of the standardized Global Protocol for Community-Scale GHG Emissions Inventories – developed jointly by the World Resources Institute, C40, and ICLEI – Local Governments for Sustainability (formerly named International Council for Local Environmental Initiatives) – is a step toward transparency and comparability between cities and over time.

Second, city governments are increasingly interested in being held globally accountable for climate change mitigation and data have played an important role in facilitating or structuring this accountability. Using the example of the

C40 city network, Gordon (2016) identifies three important sites in the politics of metrics-driven accountability. First is the use of transparency and reporting as a means for securing external recognition from investors and higher levels of government. Second is the use of transparency and reporting systems as a tool of governance and a claim to power; they help to demonstrate the effectiveness and compliance of cities. Gordon uses the example of the World Bank linking city leadership and recognition to the need for quantification, standardization, and transparency. This emphasis on quantification and standards is useful for comparability but privileges particular types and sources of knowledge about the contributions of cities to global carbon emissions. Further, as cities become more embedded in the transnational climate governance complex, the 'constituencies' to whom they are holding themselves accountable diversifies and expands beyond the local (Widerberg & Pattberg, 2017), creating the potential for tensions and contradiction in accountability metrics.

Third, in deciding what metrics are used as measures of accountability particular interests, priorities, and collective goals become embedded in the institutional infrastructure. A common theme in this work is the notion that the construction of data-driven accountability tools is itself a political act and determines to whom and for what cities are accountable. When accountability systems are shaped by political or private logics, or externally oriented, they have the potential to undermine public values and environmental outcomes. While data-centred accountability is typically conceived of as an important tool for transparency and public accessibility (Goldsmith & Crawford, 2014), the institution building around accountability and reporting schemes determines the extent to which this is true (Kitchin et al., 2015). Issues can be marginalized that are not quantifiable or included in datasets, making it difficult for constituents to hold city governments accountable for the issues they care about (van Kersbergen & van Waarden, 2004). Furthermore, data interpretation can also be subject to particular logics and rationalities, and data availability can limit the means for accountability when prohibitively expensive or not readily available (Batty, 2016).

Finally, data transparency can be used to shift accountability for climate action from the public to the private sector, which has implications for what is prioritized. Many energy use benchmarking ordinances rely on market logics and trends as key mechanisms for accountability and use transparency as a means of shifting accountability for energy conservation and GHG emissions reductions away from city hall and to building owners, buyers, and tenants. The programs make public the energy use (both absolute and relative) of the city's commercial and institutional buildings, but are rarely coupled with stringent requirements on that energy use. Rather, there is an assumption that transparency will facilitate behaviour change, and a reliance on a real estate market that rewards efficiency and environmental values.

Similarly, the metrics used in international discourses of urban climate change accountability reflect an interest in demonstrating cities as attractive sites for investment rather than the social imperatives of reducing GHG emissions. For example, highlighting efficiencies and demonstrating GHG emissions reductions can help attract investors, exemplified by a recent Price Waterhouse Coopers report that lists transparency 'as an essential factor driving the ability of cities to secure access to private capital and much-needed investment' (Gordon, 2016). In this way, accountability metrics can have more or less to do with community and local concerns and while 'city-networks like the C40 may be contesting the *who* of global climate governance, but in seeking recognition through practices of external accountability they remain firmly embedded in reproducing the prioritization of economic over environmental objectives' (Gordon, 2016).

These accountability issues also point towards the democratic elements of data in the urban climate change context. This has to do with the limited transparency in the use of urban data, such as the motivations behind utilizing certain datasets over others and the goals that are being pursued by doing so. The following section asks who is being included and excluded in this process and what the implications for data-driven climate change mitigation in cities are.

Democratic Implications of Data-Driven Climate Change Mitigation

Data-driven urban climate change mitigation has important implications for democratic urban governance. On one hand, data can empower non-state actors to act or to demand new things from the state as they gain access to new data or novel opportunities to engage in the governance of low carbon transitions. On the other hand, data can also disempower those who are not counted, who are misrepresented, or who lack access to data streams. The political nature of data means that it empowers actors unevenly, which raises essential questions of responsibility and justice for urban climate change governance.

The empowerment/disempowerment tension of data-driven climate change mitigation links to broader work interrogating the role of democracy in sustainability transitions. While the roots of sustainable development intertwine with considerations of democracy, in some places democratic struggle has been overlooked and replaced with a focus on metrics and outcomes (Stirling, 2015). In other places, scholars have connected democracy with sustainability by focusing on public participation in local environmental governance (Chu et al., 2016; Royo et al., 2014). A narrowed focus on metrics can lead to technical processes of transition with a limited scope, as opposed to broad transformation of both social practices and technologies that involve 'diverse, emergent and unruly political re-alignments' (Stirling, 2015: 54). Critical scholarship has called for renewed attention to the connections between democracy and sustainability transitions (Bäckstrand, 2006; Meadowcroft, 2009; Shove & Walker, 2007). Similarly, scholars have called for closer attention to the power relations and social justice implications of urban climate governance (Bulkeley, Castán Broto, & Edwards, 2015; Bulkeley, Edwards, & Fuller, 2014; Cohen, 2017). Considering these tensions in the literature, data-driven climate change mitigation presents opportunities to both empower new agents and engender disempowerment in several forms.

Empowering Agents

Climate change mitigation sometimes requires new information about urban infrastructure that has not been available in the past. These new streams of data can allow incumbent actors to address challenges posed by climate change. For instance, big data analytics and smart city technologies in the hands of utilities and electricity grid operators can support the increased integration of renewable energy sources (Zhou, Fu, & Yang, 2016). Data-driven climate change mitigation can also empower new agents not previously engaged in urban climate governance by creating new sources of carbon and energy use data. As we have noted, building energy use benchmarking programs have recently been established in cities like San Francisco (San Francisco, 2013) and New York City (New York City, 2014) that require large building owners to report on energy use, conduct energy audits and release the data publicly. Building owners, the public, and the private sector can access the data to try to drive building energy efficiency retrofits.

In addition, new agents may be empowered to mitigate climate change by gaining access to existing climate and energy data that has been unlocked. As an example, open data movements have called for data on energy use to be released in order to broaden access to data on patterns of energy consumption with the intention of enabling energy efficiency. Traditionally this information has been restricted to energy utilities that are usually structurally disincentivized from pursuing energy efficiency. In the United States, the Green Button Initiative responds to a 'call-to-action to provide utility customers with easy and secure access to their energy usage information in a consumer-friendly and computer-friendly format for electricity, natural gas, and water usage' (Green Button Alliance, 2017). The data can also be made available to third parties to develop technologies and applications targeting increased energy efficiency in households, although there are critiques of the technocentric approach of many of these initiatives (Bickerstaff, Hinton, & Bulkeley, 2016). Similarly, movements to open up access to urban planning data have created new ways to enable agents through opportunities like hackathons. An increasingly popular format, hackathons gather participants in a shared space to try to find ways to make big data useable or to brainstorm new ways to collect urban data (Lodato & DiSalvo, 2016).

New urban actors can also be empowered by data-driven climate change mitigation through opportunities for participation. The development of urban greenhouse gas emission inventories and local climate action plans often incorporates consultation or participation of stakeholders, including citizens, NGOs, and private sector representatives. In Oxford, an initiative launched in 2010 called Low Carbon Oxford created a city-wide programme of collaboration across private, public, and non-profit sectors. The initiative worked with 29 organizations that represented a large proportion of Oxford City's carbon emitters to both gain data about actual local emissions and to help the organizations reduce their carbon emissions (Giest, 2017). Despite the democratic openings identified in this section, however, patterns of disempowerment in data-driven urban climate change mitigation practices raise concerning implications for democratic struggle.

Disempowerment in Many Forms

Three key implications for disempowerment can be drawn from urban carbon calculation approaches that are common practice in data-driven climate change mitigation. The first is the responsibilization of individuals. It is increasingly characteristic to position the individual as the primary agent responsible for climate action (as opposed to governments or the private sector) (Fuller, 2017). Individuals are then influenced to govern their own GHG emissions by doing things like investing in energy-efficient technologies or changing their behaviour (Paterson & Stripple, 2010). Climate action is hardly unique, as the mobilization of individuals has been broadly identified as a growing pattern across environmental governance under the influence of neoliberalism (Brand, 2007). The implication of the focus on individuals, however, is the disempowerment of powerful agents in government and industry who have control over significant GHG emissions and, importantly, could influence structural drivers of urban emissions beyond the reach of the individual. More broadly, the focus on individual behaviour redirects attention from the task of tackling carbon-intensive urban development (Rice, 2014).

The second disempowerment implication of carbon calculation is the tendency to project an image of comprehensive action while, in reality, empowering only a particular subset of potential response actions. The polished appearance of quantitative inventories and climate response action plans suggest that carbon calculations are comprehensive and politically benign. In fact, a number of decisions are subjectively made about what to include and exclude in GHG emission inventories, as well as about what kinds of response actions are appropriate

(Bulkeley, 2013; Rice, 2014). In Portland, Oregon, USA, for example, the local climate response plan emphasizes energy efficiency in order to enhance the political palatability of the response strategy (Rutland & Aylett, 2008). Not only is the GHG emission impact of this strategy questionable because energy efficiency is not the same thing as reducing energy consumption, but the result of this emphasis is empowerment to do some things and not others. Citizens are not encouraged to lobby polluting industries, but, instead, are encouraged to invest in home insulation (Rutland & Aylett, 2008). As this example shows, data are often deployed in GHG inventories and local climate action plans to justify technological solutions instead of identifying the need to tackle capitalist urbanization as a cause of the problem (Rice, 2014).

Third, carbon calculations also frequently have a homogenizing effect on urban populations, with implications for justice and equity. Carbon footprints are one common way of representing responsibility for GHG emissions, but they distribute responsibility for climate action evenly across urban populations (Fuller, 2017). These carbon calculation practices homogenize urban citizens under a veneer of apolitical data even though spatial and demographic variation in GHG emissions is well documented (Jones & Kammen, 2011). As Rice (2014: 381) puts it, 'urban carbon governance erases importance aspects of social and spatial difference among carbon emitters'. For an example of the lived experience, consider Hong Kong. Despite the fact that wealth disparity in Hong Kong is one of the highest in the world, carbon footprint calculations are based on the idea of universal overconsumption (Fuller, 2017). The framing of collective responsibility leads to a collective target – a 10 per cent per capita reduction in each citizen's carbon footprint (Fuller, 2017). Fuller's (2017) research in Shek Kip Mei in Hong Kong found that 50 per cent of households had an uncomfortable indoor air temperature (above 30 degrees Celsius) because they did not have access to or could not afford air conditioning. Considering these circumstances of poverty, the idea of holding all Hong Kong citizens to a requirement to reduce emissions by 10 per cent seems unjust (Fuller, 2017). In considering the future of urban carbon calculation, therefore, there is a critical need to interrogate the 'moral and political values at the heart of such calculative devices' (Fuller, 2017: 520).

Finally, there are democratic implications tied to issues of access and control over data. There are concerning implications for democracy when private corporations have control over data that is essential to public policy and urban governance. This is an issue that has been addressed in smart cities debates. Concerns have been raised about the involvement of corporations in smart city initiatives and the resulting influence of corporate interests on urban governance (Kitchin, 2013), In addition, data related to urban climate change mitigation can have sensitive ethical and political implications (Batty, 2016; Kitchin, 2015). As we have previously stated, smart city rhetoric tends to present a technocentric and apolitical vision for cities where flows of data enable low-carbon and sustainability transitions, while issues such as 'panoptic surveillance, technocratic and corporate forms of governance, technological lock-ins, profiling and social sorting, anticipatory governance, control creep, the hollowing out of state provided services, widening inequalities and dispossession of land and livelihoods' remain untouched (Kitchin, 2015: 132).

In summary, data-driven climate change mitigation can broaden the availability of data sources that are essential ingredients for urban climate response. Related practices such as local GHG emission inventories and action plans can create democratic openings for broader community participation in urban climate governance. However, the responsibilization of individuals, the erasure of social and spatial differences in carbon emissions, and issues of data control and the corporatization of climate action have concerning implications for democracy and social justice in data-driven urban climate change mitigation.

Discussion and Conclusion

Using data for urban climate change mitigation is a complex and challenging process. Scholars have raised concerns about the political implications of data-driven mitigation, from initial data collection to its use in decision-making. Whereas collecting and incorporating larger amounts and new kinds of data may on the surface appear straightforward, it has important implications for policy-making, accountability, and democracy. For urban climate change mitigation, there are rarely mechanisms in place to encourage carbon management that meets progressive social goals (While et al., 2010). It is therefore important to consider the ways data can be used as a tool for empowerment or disempowerment. As Zook (2017: 11) warns, we 'must ever be mindful that metrics don't simply measure; in the process of deciding what is important and possible to measure, these data are simultaneously defining what cities are'. Earth System Governance scholars are playing an important role in highlighting the lack of political neutrality in the move toward data-driven climate change mitigation (e.g. Bulkeley et al., 2011; Gordon, 2016; Kramarz & Park, 2016).

In addition to ongoing critical research related to the politics of data-driven urban climate change mitigation, we highlight three particular areas of inquiry important for scholarship and practice going forward. First, future research should delve more deeply into identifying who is responsible for the various dimensions of data collection, storage, and use in cities – the actors implicated in these systems and their interests, resources, and strategies. This goes beyond the legal implications of data sharing, towards the more subtle implications of what it means to use certain types of data over others and the inclusion and exclusion that follows. As our review illustrates, the question of who has control over data remains an important consideration for democratic urban governance. For example, Kramarz and Park (2016: 2) highlight the importance of linking 'the goals of actors to the design of environmental institutions', and thereby moving toward ways of using data less as a policing resource for accountability and more for spurring environmental action. We urge future research to examine the asymmetries that exist in the data and how that affects the work of city governments and urban residents.

Second, critiques of data-driven urban climate change mitigation should begin to move toward identifying productive ways forward for cities as they navigate the emerging, complex terrain of 'big data' and 'smart cities' (Shelton, 2017). The demand for data-driven decision-making is not likely to lesson, and there are genuine advantages to be had from the availability of new and more information. Research can contribute to understanding when and how more data contributes to better decision-making, progressive practices for data collection and use in cities, and the means by which the data-driven city movement can be used to enhance social and economic equity. For example, McLaren and Agyemen argue for 'sharing cities' as mechanisms for forwarding the goals of just sustainability (McLaren & Agyeman, 2015). The data-driven movement holds potential and promise for cities if harnessed for progressive aims.

Third, systematic comparisons between cities can help highlight how data-driven decision-making reflects or reproduces existing political and institutional structures (Luque-Ayala & Marvin, 2015). For example, Raven et al. (2017) found that the ways cities experiment with 'smart city' initiatives are heavily influenced by national and subnational governing logics and institutional norms. As cities continue to engage in data-driven decision-making for climate change mitigation there will be increasing opportunities to learn from their approaches and develop meaningful partnerships for addressing the social and political challenges cities will encounter.

References

Anguelovsky, I., & Carmin, J. (2011). Something borrowed, everything new: Innovation and institutionalization in urban climate governance. *Current Opinion in Environmental Sustainability*, 3(3): 169–175.

Bäckstrand, K. (2006). Democratizing global environmental governance? Stakeholder democracy after the World Summit on Sustainable Development. *European Journal of International Relations*, 12(4): 467–498. https://doi.org/10.1177/1354066106069321

Batty, Michael. (2016). Big data and the city. *Built Environment*, 42(3): 321–337. DOI:10.2148/benv.42.3.321.

Bickerstaff, K., Hinton, E., & Bulkeley, H. (2016). Decarbonisation at home: The contingent politics of experimental domestic energy technologies. *Environment and Planning A*, 48(10): 2006–2025. http://doi.org/10.1177/0308518X16653403

Bloomberg Finance. (2018). Approach. https://www.bloomberg.com/impact/approach/ (accessed 15 November 2018).

Brand, P. (2007). Green subjection: The politics of neoliberal urban environmental management. *International Journal of Urban and Regional Research*, 31(3): 616–632. http://doi.org/10.1111/j.1468-2427.2007.00748.x

Bui, L. (November 2014). Remote controls, switchboards, dashboards, war rooms: The cybernetics DNA of smart cities. Medium, 2 November 2014. https://medium.com/@dangerbui/remote-controls-and-dashboards-is-the-responsive-city-an-authoritarian-w-7617e8a83d9

Bulkeley, Harriet, Broto, Vanesa Castán, Hodson, Michael, and Marvin, Simon (2011). *Cities and Low Carbon Transitions*. Oxford: Routledge.

Bulkeley, H. (2013). *Cities and Climate Change*. New York: Routledge.

Bulkeley, H., Castán Broto, V., & Edwards, G. A. S. (2015). *An Urban Politics of Climate Change: Experimentation and the Governing of Socio-Technical Transitions*. London: Routledge.

Bulkeley, H., Edwards, G. A. S., & Fuller, S. (2014). Contesting climate justice in the city: Examining politics and practice in urban climate change experiments. *Global Environmental Change*, 25: 31–40. http://doi.org/10.1016/j.gloenvcha.2014.01.009

Bulkeley, H., McGuirk, P. M., & Dowling, R. (2016). Making a smart city for the smart grid? The urban material politics of actualizing smart electricity networks. *Environment and Planning A*, 48(9): 1709–1726. DOI:10.1177/0308518X16648152.

Chu, E., Anguelovski, I., & Carmin, J. (2016). Inclusive approaches to urban climate adaptation planning and implementation in the Global South. *Climate Policy*, 16(3): 372–392. DOI:10.1080/14693062.2015.1019822.

Cohen, D. A. (2017). The other low-carbon protagonists: Poor people's movements and climate politics in São Paulo. In M. Greenberg & P. Lewis (eds.), *The City Is the Factory: New Solidarities and Spacial Strategies in an Urban Age*. Ithaca, NY: Cornell University Press, 140–157.

Cotton, J. (September 2015). Christchurch's Sensing City project shelved. Idealog, 4 September 2015. http://idealog.co.nz/tech/2015/09/christchurchs-sensing-city-project-shelved (accessed 15 November 2018).

Cox, Matt, Brown, Marilyn A., & Sun, Xiaojing. (2013). Energy benchmarking of commercial buildings: A low-cost pathway toward urban sustainability. *Environmental Research Letters*, 8(3): 035018. DOI:10.1088/1748-9326/8/3/035018.

de Waal, M. (2017). A city is not a galaxy: Understanding the city through urban data. In R. Kitchin, T. P. Lauriault, & G. McArdle (eds.), *Data and the City*. New York: Routledge, 17–30.

Edelenbos, J., Hirzalla, F., van Zoonen, L., et al. (2017). Governing the complexity of smart data cities: Setting a research agenda. In M. P. Rodriguez Bolivar (ed.), *Smart Technologies for Smart Governments: Transparency, Efficiency and Organizational Issues*. Cham, Switzerland: Springer International, 35–54.

FitzGibbon, J., & Mensah, K. O. (2012). Climate change as a wicked problem: An evaluation of the institutional context for rural water management in Ghana. *SAGE Open*, 1–14. DOI:10.1177/2158244012448487.

French, S., Barchers, C., & Zhang, W. (2015). Moving beyond operations: Leveraging big data for urban planning decisions. CUPUM paper. http://web.mit.edu/cron/project/

CUPUM2015/proceedings/Content/pss/194_french_h.pdf (accessed 15 November 2018).

Fuller, S. (2017). Configuring climate responsibility in the city: Carbon footprints and climate justice in Hong Kong. *Area*, 37: 23–27. http://doi.org/10.1111/area.12341

Giest, S. (2017). Big data analytics for mitigating carbon emissions in smart cities: Opportunities and challenges. *European Planning Studies*, 25(6): 941–957. DOI:10.1080/09654313.2017.1294149.

Goldsmith, Steven, and Crawford, Susan. (2014). *The Responsive City: Engaging Communities through Data-Smart Governance*. San Francisco: Jossey-Bass.

Gordon, David J. (2016). Politics of accountability in networked urban climate governance. *Global Environmental Politics*. www.mitpressjournals.org/doi/abs/10.1162/GLEP_a_00357

Graham, S. (2005). Software-sorted geographies. *Progress in Human Geography*, 29(5): 562–580.

Green Button Alliance. (2017). Green button data. Green Button Alliance. www.greenbuttondata.org/ (accessed 15 November 2018).

Greenfield, A. (2013). *Against the Smart City*. New York: Do Publications.

Head, B. W., & Halford, J. (2015). Wicked problems: Implications for public policy and management. *Administration & Society*, 47(6): 711–739.

Höchtl, J., Parycek, P., & Schöllhammer, R. (2016). Big data in the policy cycle: Policy decision making in the digital era. *Journal of Organizational Computing and Electronic Commerce*, 26(1–2): 147–169. DOI:10.1080/10919392.2015.1125187.

Howlett, M. (2009). Policy analytical capacity and evidence-based policy-making: Lessons from Canada. *Canadian Public Administration*, 52(2): 153–175. DOI:10.1111/j.1754-7121.2009.00070_1.x.

Innes, J. E., & Booher, D. E. (March 2003). The impact of collaborative planning on governance capacity. Working Paper 2003, 03, Institute of Urban and Regional Development, University of California, Berkeley. http://escholarship.org/uc/item/98k72547 (accessed 15 November 2018).

Jones, C. M., & Kammen, D. M. (2011). Quantifying carbon footprint reduction opportunities for U.S. households and communities. *Environmental Science & Technology*, 45 (9): 4088–4095. http://doi.org/10.1021/es102221

Kitchin, R. (2013). The real-time city? Big data and smart urbanism. *GeoJournal*, 79(1): 1–14. http://doi.org/10.1007/s10708-013-9516-8

Kitchin, R. (2015). Making sense of smart cities: Addressing present shortcomings. *Cambridge Journal of Regions, Economy and Society*, 8(1): 131–136.

Kitchin, Rob, Lauriault, Tracey P., and McArdle, Gavin. (2015). Knowing and governing cities through urban indicators, city benchmarking and real-time dashboards. *Regional Studies, Regional Science*, 2(1): 6–28.

Kitchin, R. (2014). *The Data Revolution: Big Data, Open Data, Data Infrastructures and Their Consequences*. London: SAGE.

Kitchin, R. (2017). Data-driven urbanism. In R. Kitchin, T. P. Lauriault, & G. McArdle (eds.), *Data and the City*. New York: Routledge, 44–56.

Kitchin, R., Lauriault T. P., & McArdle, G. (2017). Data and the city. In R. Kitchin, T. P. Lauriault, & G. McArdle (eds.), *Data and the City*. New York: Routledge, 1–14.

Kitchin, R., Maalsen, S., & McArdle, G. (2016). The praxis and politics of building urban dashboards. *Geoforum*, 77: 93–101.

Kitchin, R., & McArdle, G. (2017). Urban data and city dashboards: Six key issues. In R. Kitchin, T. P. Lauriault, & McArdle, G. (eds.), *Data and the City*. New York: Routledge, 111–126.

Knox-Hayes, J., & Levy, David L. (2011). The politics of carbon disclosure as climate governance. *Strategic Organization*, 9(1): 91–99. DOI:10.1177/1476127010395066.

Kourtit, K., Nijkamp, P., & Arribas, D. (2012). Smart cities in perspective: A comparative European study by means of self-organizing maps. *Innovation: The European Journal of Social Science Research*, 25(2): 229–246.

Kramarz, Teresa, & Park, Susan. (2016). Accountability in global environmental governance: A meaningful tool for action? *Global Environmental Politics*, 16(2): 1–21.

Kramarz, Teresa, & Park, Susan. (2017). Introduction: The politics of environmental accountability. *Review of Policy Research*, 34(1): 4–9. DOI:10.1111/ropr.12223.

Laney, D. (2001). 3D data management: Controlling data volume, velocity, and variety. Gartner Group. http://blogs.gartner.com/doug-laney/files/2012/01/ad949-3D-Data-Management-Controlling-Data-Volume-Velocity-and-Variety.pdf (accessed 15 November 2018).

Larson, K., White, D., Gober, P., & Wutich, A. (2015). Decision-making under uncertainty for water sustainability and urban climate change adaptation. *Sustainability*, 7: 14761–14784. DOI:10.3390/su71114761.

Levin, K., Cashore, B., Bernstein, S., & Auld, G. (2012). Overcoming the tragedy of super wicked problems: Constraining our future selves to ameliorate global climate change. *Policy Sciences*, 45(2): 123–152.

Lodato, T. J., & DiSalvo, C. (2016). Issue-oriented hackathons as material participation. *New Media & Society*, 18(4): 539–557. http://doi.org/10.1177/1461444816629467

Luque-Ayala, Andrés, & Marvin, Simon. (2015). Developing a critical understanding of smart urbanism? *Urban Studies*, 52(12): 2105–2116.

Marvin, Simon, Luque-Ayala, Andres, & McFarlane, Colin, eds. (2016). *Smart Urbanism: Utopian Vision or False Dawn?* London: Routledge.

Mattern, Sara. (2013). Municipal energy benchmarking legislation for commercial buildings: You can't manage what you don't measure. *Boston College Environmental Affairs Law Review*, 40(2): 487.

McLaren, Duncan, & Agyeman, Julian. (2015). *Sharing Cities: A Case for Truly Smart and Sustainable Cities*. Cambridge, MA: MIT Press.

Meadowcroft, J. (2009). What about the politics? Sustainable development, transition management, and long term energy transitions. *Policy Sciences*, 42(4): 323–340. http://doi.org/10.1007/s11077-009-9097-z

Meijer, A., & Bolivar, M. (2016). Governing the smart city: A review of the literature on smart urban governance. *International Review of Administrative Sciences*, 82(2): 392–408. DOI:10.1177/0020852314564308.

Morozov, E. (2013). *To Save Everything, Click Here: Technology, Solutionism and the Urge to Fix Problems That Don't Exist*. London: Allen Lane.

New York City. (2014). *One City: Built to Last*. New York: New York City.

Nutley, S. M., Walter, I., & Davies, H. T. O. (2007). *Using Evidence: How Research Can Inform Public Services*. Bristol: Policy Press.

Paterson, M., & Stripple, J. (2010). My Space: Governing individuals' carbon emissions. *Environment and Planning D: Society and Space*, 28(2): 341–362. http://doi.org/10.1068/d4109

Pawson, R. (2006). *Evidence-Based Policy: A Realist Perspective*. London: SAGE.

Raven, Rob, Sengers, Frans, Spaeth, Philipp, Xie, Linjun, Cheshmehzangi, Ali, & de Jong, Martin. (2017). Urban experimentation and institutional arrangements. *European Planning Studies*, 27(2): 258–281. https://doi.org/10.1080/09654313.2017.1393047

Rice, J. L. (2014). An urban political ecology of climate change governance. *Geography Compass*, 8(6): 381–394. http://doi.org/10.1111/gec3.12134

Rittel, H., & Webber, M. (1973). Dilemmas in a general theory of planning. *Policy Sciences*, 4: 155–169.

Romero-Lankao, P. (2012). Governing carbon and climate in the cities: An overview of policy and planning challenges and options. *European Planning Studies*, 20(1): 7–26. DOI:10.1080/09654313.2011.638496.

Royo, S., Yetano, A., & Acerete, B. (2014). E-participation and environmental protection: Are local governments really committed? *Public Administration Review*, 74(1): 87–98. DOI:10.1111/puar.12156.

Rutland, T., & Aylett, A. (2008). The work of policy: Actor networks, governmentality, and local action on climate change in Portland, Oregon. *Environment and Planning D: Society and Space*, 26(4): 627–646. http://doi.org/10.1068/d6907

San Francisco. (2013). Climate Action Strategy 2013 Update, 1–73.

Sanderson, I. (2006). Complexity, 'practical rationality' and evidence-based policy making. *Policy & Politics*, 34(1): 115–132. DOI:10.1332/030557306775212188.

Shelton, Taylor. (2017). The urban geographical imagination in the age of big data. *Big Data & Society*, 4(1): 2053951716665129.

Shove, E., & Walker, G. (2007). CAUTION! Transitions ahead: Politics, practice, and sustainable transition management. *Environment and Planning A: Economy and Space*, 39(4): 763–770. http://doi.org/10.1068/a39310

Stirling, A. (2015). Emancipating transformations: From controlling "the transition" to culturing plural radical progress. In I. Scoones, M. Leach, & P. Newell (eds.), *The Politics of Green Transformations*. New York: Routledge, 54–67.

Tomer, A., & Shivaram, R. (July 2017). Modernizing government's approach to transportation and land use data: Challenges and opportunities. Brookings, 20 July 2017. www.brookings.edu/research/modernizing-approach-to-data/ (accessed 15 November 2018).

Trenberth, K. (2010). More knowledge, less certainty. *Nature Reports Climate Change*, 4: 20–21. DOI:10.1038/climate.2010.06.

van Kersbergen, C. J., & van Waarden, F. (2004). 'Governance' as a bridge between disciplines: Cross-disciplinary inspiration regarding shifts in governance and problems of governability, accountability and legitimacy. *European Journal of Political Research*, 43(2): 143–171. DOI:10.1111/j.1475-6765.2004.00149.x.

While, A., Jonas, A. E. G., & Gibbs, D. (2010). From sustainable development to carbon control: Eco-state restructuring and the politics of urban and regional development. *Transactions of the Institute of British Geographers*, 35: 76–93.

Widerberg, Oscar, & Pattberg, Philipp. (2017). Accountability challenges in the transnational regime complex for climate change. *Review of Policy Research*, 34(1): 68–87. https://doi.org/10.1111/ropr.12217

Vanolo, A. (2014). Smartmentality: The smart city as disciplinary strategy. *Urban Studies*, 51(5): 883–898.

Zhou, K., Fu, C., & Yang, S. (2016). Big data driven smart energy management: From big data to big insights. *Renewable and Sustainable Energy Reviews*, 56(c): 215–225. http://doi.org/10.1016/j.rser.2015.11.050

Zook, Matthew. (2017). Crowd-sourcing the smart city: Using big geosocial media metrics in urban governance. *Big Data & Society*, 4(1): 2053951717694384.

8

Urban Planning for Sustainability and Justice

Lessons from Urban Agriculture

FRANÇOIS MANCEBO AND CHIARA CERTOMÀ

Urban Transitions to Sustainability: What Exactly Is It About?

Sustainability and justice have become two alleged priorities of urban policies, especially those addressing climate change issues. Both of these two priorities are challenging to achieve in urban policies. The present chapter focuses on how adaptive planning and participative governance can offer the opportunity to plan for sustainability and justice at once through observations in the field of urban agriculture.

The primary obstacle for a planner when addressing simultaneously sustainability and justice is that urban planning generates inherently wicked problems: namely, difficult to define, unpredictable, and defying rational decision-making (Rittel & Webber, 1973). As long ago as in the 1960s Jane Jacobs already observed that urban matters were neither rational problems waiting for solutions, nor a complete chaos: 'Problems which involve dealing simultaneously with a sizeable number of factors which are interrelated into an organic whole' (Jacobs, 1961: 432). The possibility of scientific, social and political consensus on the course of an action is unlikely due to conflicting interpretations of what the real problem is and what its causes are, which vary considerably with the different values and interests of the actors (Fischer, 2000).

Today, many of the wicked problems that cities face are related to sustainability issues (Scheffer et al., 2009). Sustainability exacerbates the wicked side of planning since sustainable development is not only about science but also about values, which may differ greatly between cultures and over time (Christen & Schmidt, 2012; Leiserowitz et al., 2006). This last decade, more and more planners began to realize – when trying to implement sustainability policies – that new urban structures and behaviours emerged constantly from wicked problems, in an unpredictable way (Allen, 1997). It became obvious that discontinuous and chaotic change reigned everywhere in urban areas (Batty, 2008). It had always been the case: Just consider the great fires and floods of previous centuries in Paris or London, the consequence of Middle Ages' plagues on the cities and later the industrial cities' metamorphosis in the nineteenth century. Yes, urban futures were always chaotic, but the difference today is that there are tools to deal with uncertainty.

One of these tools is adaptive planning. Adaptive planning issues have been addressed these last years within Earth System Governance research, especially with the special issue 'Earth System Governance – Task Force Initiative on Sustainability Science' published recently in the review *Challenges in Sustainability* (Isgren et al., 2017). The very notion of adaptiveness was introduced in 2009's Earth System Governance Science Plan. A key idea of adaptive planning is that when too much emphasis is put on problem identification and solution, it usually ends in unintended negative outcomes (Verweij & Thompson, 2007). As a matter of consequence, urban planning should be less about how to find solutions to predetermined problems than understanding the dynamics that give rise to desirable and undesirable phenomena: planning has to move from a prescriptive activity to a process of learning. It entails collaborative process engaging communities, professionals, and other stakeholders with urban planners. Workshops and joint fact-finding and public forums may help fostering synergistic urban lifestyles that are desirable, attainable, maintainable, and reproducible – realizing what is usually called 'meta design planning' (Justus & Taylor, 2011).

But – as part of adaptive planning – collaborative action is anything but obvious. The greatest and more general difficulty is lack of legitimacy for the process itself and for its outcomes (Lang et al., 2012) as typified by the following case: In the province of Limburg (Netherlands) the results of a study, whose objective was to measure and develop sustainability planning, has never been adopted by the local and regional authorities who sponsored it. The reason given was that the partners of the civil society who worked within the research team 'had no political mandate for defining sustainable development in this regional context' (Van Zeijl-Rozema & Martens, 2011). Such a situation is not uncommon: When trying to generate knowledge for collective action, the process and its outcomes often interfere with legitimized procedures and official politics (Scholz, 2011).

There is another important issue to adaptive planning: integrating bottom-up processes of knowledge and data collection and top-down agency (Brenman & Sanchez, 2012). This issue can be embodied in two questions: How can a planner know enough about the lives of local people to propose the best possible policies? How is a community motivated – or not – to collect its local information and communicate it in a way that can help planners? The example of the public water points in the city of Pune (India), developed by Luis Bettencourt, raises the following issue: How is it possible for a planner to determine how many public water points should be created in a neighbourhood (Bettencourt, 2015)? From the inhabitants' point of view, short distance and easy maintenance are essential, as well as minimal waiting time, which calls for a large number of points forming a dense network. Such a choice presents a collateral interest: Smaller groups use every point, which fosters a stronger sense of responsibility. But how does the planner know how many points are not too many? He has to learn it from the communities themselves. The inhabitants are the only ones who know the real limits – not the administrative limits – of the communities and of the neighbourhoods. But they will give the information only if they perceive that it is in their best interest and if they feel they will have a seat at the decision-making table. This type of urban planning entails trust, as well as knowledge issues.

Addressing Sustainability and Justice through Adaptive Planning and Collaborative Action

From a normative perspective, everyone concerned by sustainability issues should be involved in the process of decision-making (Gibson, 2006). From a strategic perspective, people in the general population have values and knowledge that are out of reach of experts, scientists, or elected representatives, and may prove essential to effective sustainability decision-making (Fischer, 2000). These two complementary standpoints indicate that sustainability resonates strongly with the notion of participatory justice, and not only distributive justice or access (Klinsky & Golub, 2016). Maximizing wide-scale involvement in urban planning improves justice, which is to be expected because it is impossible to define justice independently from its social context (Miller, 2002; Walzer, 1983). Thus, the present section aims at showing how collaborative action can be relevant when coping simultaneously with sustainability and justice.

Embedding all concerned people, with the different needs and values in a collaborative action has a big consequence: No panaceas exist, to use the words of Elinor Ostrom (Ostrom & Cox, 2010). The solutions always depend on the characteristics of the local communities, which means that a place-based approach is necessary (Daily et al., 2009). It is therefore essential to determine locally what is a good environment for the communities involved: one in which the improvement of environmental conditions *stricto sensu* (water quality, air, biodiversity, energy, etc.) will lead to improved living conditions; one in which technical devices and ecological processes will lead to new lifestyles. And it is all but obvious.

For example, one among the many challenges of sustainability should be making a better use of what is already there (Mancebo, 2015). Sustainable cities could thus be nicknamed 'recyclable cities' in the sense that they are supposed to recycle their urban fabric and their urban functions without going through phases of obsolescence with brownfield land and degraded neighbourhoods, and without squandering soils (Swart et al., 2003). Unfortunately, usually this is not what happens. Mayors, representatives, and more generally elected officials take great pleasure in showcasing constructions and they love them 'brand new'. They are less interested in the urban design, which is more important to create a real sustainable city but, of course, harder to implement and less profitable as an electoral issue. 'Exemplary' buildings and devices are often favoured to the detriment of more holistic approaches, such as active land management and transformation of the urban fabric (differential densification, restructuring urban cores, etc.). In many cases, vegetation, green technologies, and exterior wood facing camouflage very classical housing estates totally disconnected from their surroundings. There is another issue concerning social justice in these cases: Wouldn't public money have been spent more efficiently invested to reduce actual environmental disparities between areas, bettering underprivileged districts where the environmental conditions are already pretty bad, as in the metropolitan region of Paris? In the metropolitan region of Paris 50 per cent of the places with degraded environment are also socially deprived; symmetrically, nearly 50 per cent of those with good environmental conditions are wealthy areas (Bigot, 2009). If we try to consider what the main factor for such a distribution is, the attractiveness of the *communes* with a nice environment appears less decisive than the avoidance of the nuisances of those with a poor one. What is interesting here is that the residential choice is motivated by the rejection of environmental degradation rather than the attractiveness of environmental amenities (nature, silence, air and water quality, etc.) (Gueymard & Faburel, 2008). Thus urban sustainability policies should focus on an inclusive approach, rather than to keep on creating 'attractive' green housing spots haphazardly.

How to do so? This is accomplished by ensuring that the definition and implementation of these policies are a collective concern involving all the inhabitants, and not only the elected officials, developers, and city planners. Decision-making processes should be fundamentally a matter of collective decision (Fischhoff et al., 1981). The point is building trust so as to develop participative governance. Adaptive planning for urban futures is a matter of collective ownership and participatory joint-construction (Andrews, 2002): coproducing collective decision through interactions in which each and every actor with its own interests interacts with heterogeneous knowledge producers, including local communities and individuals able to form self-determined user associations in the continuity of Elinor Ostrom's work (Ostrom, 1998). There are three main obstacles as reported during the Earth System Governance event 'Urban Transitions to Sustainability' (Mancebo & Sachs, 2015):

- First, it is difficult to encompass all the actors (regional and local authorities, non-market institutions, non-governmental organizations (NGOs), private companies, local storekeepers, unions and chambers, landowners, etc.).
- Second, there is the question of how to take into account in the process the micro-decisions made by individuals and households, which have an indirect but strong influence on collective decisions. They are shaped by the moment and the economic status of the persons: Depending on whether they feel (or are really) poor or not, they will not make the same choice if they are placed in the alternative of eating properly or going to the theatre, thermally insulating their house or paying their bills. Ostentatious choices also play a large role (Frank, 1999).
- Third, all the actors have to consider the others as legitimate partners, and the process of co-construction itself as satisfying the criteria of legitimacy, the achievement of which is all but evident (Mollinga, 2008). How can one convince each member that the discussion is not biased in support of another member's agenda?

All this means that to develop new forms of participative governance, it is crucial to understand how the population and the institutions respond to and resist changing. How heterogeneous actors may address both sustainability and justice in the practice of urban planning. To find answers to these two questions let's focus on urban agriculture.

Indeed urban agriculture projects have been mushrooming since the end of the twentieth century. They help reshaping urban landscapes and even the whole urban fabric, experimenting with alternatives to the traditional urban life and sometimes creating new commons, and bringing people together. Within a city, farmers, gardeners, and their neighbours share more than just fence lines. These initiatives typify urban transitions to sustainability as well as the emergence of new forms of adaptive planning. Urban agriculture lends particularly well to long-lasting urban policies, especially those turning environmental 'bads' – such as brownfields and wastelands – into environmental 'goods' and urban amenities. Urban agriculture in interstitial abandoned urban areas may be one of cities' main seedbeds of creative innovation. Indeed it makes great sense to consider urban agriculture when trying to address the wicked problems generated by the sustainability and justice–justice nexus (Hou, 2010; Ioannou et al., 2016).

Insights from Urban Agriculture Initiatives

Urban agriculture initiatives provide clear examples of how emerging networks of heterogeneous social actors can tackle with the sustainability–justice nexus by advancing an adaptive approach to planning. In responding to some of the most pressing urban challenges (most notably the climate change and urban ecology-related issues) (Bulkeley & Betsill, 2005; Satterthwaite, 2007), urban agriculture shows how social innovation, fuelled by genuine participative processes, can create trust relations between different actors and can mobilize the creative potential of local inhabitants via new governance configurations. These activate, in their turn, issue-oriented networks spanning from the very local to the global level on the basis of similar interests and mode of agency (Sassen, 2006).

The term 'urban agriculture' is here adopted to refer to a broad range of projects (including those initiatives often referred to as urban gardening) that focus on growing edible and non-edible plants in both public (vacant lots, abandoned interstitial areas, flower beds, traffic islands, etc.) and private (terraces, roofs, indoor gardens, farms, etc.) areas of the city for the benefit of the public (COST Action TD1106, 2016; COST Action TU1201, 2016).

Urban agriculture practice and aims are largely context-based, as different cultural and political contexts, ownership and planning regimes, socio-agricultural traditions, and administrative regulations generate peculiar forms of intervention (see 'Case Studies', COST Action TU1201, 2016). However, urban agriculture generally takes the form of allotment gardens, collective gardens (also named 'community' or 'shared gardens'), or guerrilla/street gardening (Adams et al., 2015; Reynolds, 2008). Not all the stakeholders involved in urban agriculture initiatives are motivated by similar visions of city development, nor do they necessarily share similar values. In some cases the interest for making the city more attractive for people to live in urged administrations to provide environmental amenities in the forms of quasi-private allotment gardens located in large peripheral areas of the city, as in most Eastern and Northern European countries (see 'Case Studies', COST Action TU1201, 2016). In other cases where the pressure of urbanization and population density is high, and the availability of public land is low (as it is often the case in Mediterranean European countries) citizens often take the lead in greening brownfields or mismanaged parks with the aim of mitigating pollution, noise, lack of public green spaces, criminality, and socio-environmental negativities in general (Latkowska, 2015).

Urban agriculture is aimed at reaching different goals generally related to the multiple aspects of urban sustainability by positively influencing the environmental and social quality of city space and people's life (Wakefield et al., 2007), or by promoting environmental commitment (Certomà, 2011; Hou et al., 2009; Miller, 2005). Attention towards social concerns in urban agriculture recently increased, and several contributions documented the role of gardening or farming the city as a means for helping social disadvantage (Agyeman & Erickson, 2012; Emmett, 2011), for community-building (Beckie & Bogdan, 2010; Been & Voicu, 2006), and for involving marginalized social groups (Ferris et al., 2001; Flachs, 2010; Tracey, 2007).

Among the other goals, tackling with different forms of (in)justice in the city is gaining momentum (Certomà & Tornaghi, 2015; Reynolds, 2014). When urban agriculture engaged with justice, this last assumes a clear spatial and material character because through the physical disposition of living beings and non-living things urban agriculture contrasts the less visible (and sometime obscure) side of urban planning; and, by making gardening space available, inclusive and open to collaborative initiatives, it articulates forms of resistance and alternatives to current conditions of injustices. Specifically, scholarly investigations show that urban agriculture has the potential to address a number of spatial inequalities by contrasting the erasure of public spaces, the decrease of social cohesion, and fragmentation of solidarity links and to reduce disaffection towards places and social community (Certomà & Notteboom, 2017; Hou, 2010). These positive effects in terms of social justice are attained by both investing time and energies in the planning process and in the management of the garden itself, and they justify this chapter's interest on urban agriculture as one of cities' main seedbeds of creative innovation and transitions towards sustainability and justice (Ioannou et al., 2016; Kurtz, 2001; Pagano & Bowman, 2000).

The urban agriculture movement, in fact, radically changed both citizens' and administrations' understanding of green urban space by turning this last from a chief expression of social ordering and control (Moret, 2004) to the locus for people's empowerment and creative potential expression. Since the end of the twentieth century, growing vegetables and flowers in the city space became a common practice for complementing socio-economically disadvantaged classes' daily diet in the expanding industrial cities (Crouch & Ward, 1997; Seghers & Van Molle, 2007), for supporting the impoverished economies in wartime (McKay, 2011), and for attempting new social utopian experiments (Howard, 1902). It was only with the 1970s' rise of the social justice movement, which explicitly recognized that those who have the power to plan the city also have the power to create or perpetuate justice or injustice (Harvey, 1990; Lefebvre, 1991; Soja, 1989), that urban public space, including green space, became the object of social contestation. Since then, the land itself for urban agriculture initiatives became not merely a concession granted by local administration for charity purposes, but rather the object of socio-political contestations focusing on issues of space ownership, distribution, and use (Amin & Thrift, 2002; Soja, 2000).

Today, cultivating in the city is recognized as the expression of a radically new relationship between citizens and their city (Eizenberg, 2012; Lamborn & Weinberg, 1999).

Nevertheless, it is worth mentioning some critical positions understanding urban agriculture initiatives as manifestations of individual and quasi-autarkic citizens' actions (Pudup, 2008; Weisman, 2009), able to determine controversies and injustices, including new forms of enclosures or gentrification, when handing over public space to a limited group of citizens (Tornaghi, 2014). Again, they signal the opportunistic involvement of citizens who deresponsabilize administration in restoring derelict areas of no interest for private investors (Smith & Kurtz, 2003), as in the *Kinderbaurnhof Mauerplatz Kreuzberg* case, Berlin, in the early 1980s (Rosol, 2012). Other scholars further point out that analysis of the social composition of urban agriculture activists shows that, in several cases, socio-egalitarian aims have been pioneered by middle-income people, rather than those from the working class, as happened in *DeTunjes* project, Ghent (Certomà & Notteboom, 2017).

Despite the controversies on the political interpretation of the phenomenon, however, it is broadly recognized that gardeners and farmers' engagement in urban agriculture initiatives is addressing wicked problems in which different values materialize and often conflate via the production of an effective transition towards a just and sustainable city. It is particularly interesting how the two goals are combined to respond and resist the current challenges via new forms of adaptive planning able to address the wicked problems of the sustainability–justice nexus.

Moving from Prescriptive to Adaptive Planning

In tackling the multidimensional character of urban sustainability challenges, which requires a balance between environmental protection measures, social cohesion, and the provision of justice (IPCC, 2007; U.N.-HABITAT, 2010), special attention is now devoted to the adoption of participatory and transparent approaches (Haughton & Hunter, 2003; Larsen et al., 2011; Pearsall & Pierce, 2010). In this context, urban sustainability planning necessarily requires shifting from a prescriptive activity to a process of learning. The resulting adaptive planning process is in fact the most appropriate to deal with the plethora of different issues emerging from the combination of social justice and sustainability. This requires involving citizens in decisions that affect them via an active engagement in the city design rather than via a mere consultation on already existing proposals; and it also modifies the rules of the eternal game between how the authorities – whatever their form – try to shape the social fabric, and how the social fabric impacts on the authorities, through deception or force, confrontation or bargaining (Mancebo, 2016).

Urban agriculture initiatives show us how transition to adaptive planning is possible, and how this allows encompassing all actors, considering micro-decisions, and creating trust and social cohesion. In particular, our investigation of urban agriculture illustrates that both understandings and practices of planning can change over time based on learning processes, and that co-creative experiences can work as proactive tools for accessing and distributing environmental benefits.

A number of cases worldwide (Costa et al., 2016; Drilling et al., 2016) demonstrate that urban agriculture initiatives often advance a form of adaptive planning that is focused on the reinvention of urban public spaces via the mediation of farmers, gardeners, neighbours, administration, and other interested social actors (Certomà & Notteboom, 2017). The innovative character of this adaptive planning resides in its being a collective reinterpretation and re-elaboration of the society and space relationship via grassroots-based planning practices (Roy, 2005). These practices are able to reinterpret urban void or abandoned interstitial areas even in the absence of legal definition, regulation, or funds provided by the public or private sector (Jiménez, 2014; Vestbro, 2013); and – while distancing from top-down planning practices (Mitchell, 2003) – they also seek legitimation and dialogue with traditional planning authorities (Donovan, 2008). Such an adaptive planning approach is characterized by the fast and broad creation of horizontal and vertical actor networks which significantly differ from traditional social aggregation experiences (Humphreys, 2010; Juris, 2012) and make the entire process genuinely participative because it is inspired and (initially) led mainly by heterogeneous associations of stakeholders interested in the creation of new spaces for social aggregation and for the creation of ecological networks in the city.

This is, for instance, the case of *Parco delle Energie* in Rome, where grassroots groups (including neighbourhood associations, urban farmers, and cultural associations) cleaned up, restored, and established allotments in a quasi-abandoned public park. The role of this urban agriculture initiative was crucial in catalysing urban informality practices through an emerging and innovative governance process that revitalized the entire park and the neighbourhood. The *Parco delle Energie* quickly became a hotspot for ecosystem services generation (such as clean air, green amenities, ecological corridors, fresh food in a most densely populated area of the city) and for enhancing social cohesion. The allotments and the surrounding restored the park as host, in fact, of multicultural initiatives, a popular museum of modern history, playing areas for children, strolling areas for elderly people, and open-air sport facilities. It is also the case of *La Fournillière*, a former squatted wasteland of more than 3 ha in the city of Nantes turned into a particularly charming and unusually large urban farming land that one can reach only by walking (Mancebo, 2016). In the mid-1970s gardeners progressively squatted this wasteland: There were more than 70 illegal squatting gardeners at the beginning of the 1990s, when Nante's city council finally decided to develop a park on the wasteland. Something unusual happened then. The gardeners spontaneously united their forces and organized to impose their views to the municipality. Rather than making demands and organizing protests, they decided to draw out an in-depth report on the actual situation at *La Fournillière*, providing an exact overview of the long work of clearing and planting that they had done as well as the public goods they had created; it illustrated the social and ecological value of these gardens for the whole city. Finally, gardeners claimed that they wanted to be decision-making partners in the project and have a seat at the decision-making table. At the end of a long process of negotiations – and against all odds – the city council decided to support the gardeners' alternative project now supported by the local community.

These cases exemplify the redefinition of traditional planning and management processes of public parks because the administration – unable to maintain the area subject to the increasing urbanization pressure, and inspired by the activism of citizens – proposed to establish mutually profitably agreements towards a collaborative governance protocol. Very often the adaptive planning advanced by urban agriculture initiatives requires the emergence of a new form of governance, which entails the restructuring of power relationships between community associations, public administrations, and private companies in the process of public space shaping. Examples of fruitful collaborations inspired by urban agriculture initiatives between public and private subjects are copious in Europe, with the case of the *Federación Regional de Asociaciones Vecinales* [Regional Federation of Neighborhood Associations] in Madrid being a most interesting one as most of the allotments and surrounding green areas have been cooperatively planned by citizens and local authorities (Ioannou et al., 2016).

Such a complex planning approach is generated by cross-scales network configurations where the strong presence of neighbourhood groups and citizens' organizations is combined with the administration's willingness to deal in common both responsibility and power (Gerometta et al., 2005; Moulaert et al., 2007). A clear example is provided by *De Site* community garden in Ghent, whose realization in a polluted and formerly industrial area of the city, largely populated by scarcely integrated and mostly unemployed migrant people, was possible thanks to the cooperation and the continuous shifting of power among city administration, neighbourhood groups, and a private real estate company. It is not unusual to see urban agriculture projects, originally started by community activists, evolving and entering a process of negotiation and co-management with official planners and landowners, as was the case for the *Ökotop Heerdt* in Düsseldorf (Sondermann, 2014), which evolved from a radical activist ecological garden project into an administration-led project for experimenting participative planning. The adaptive planning approach is often advanced by local administrations themselves that promote urban agriculture projects as tools for both marketing the city and creating equitable and healthy conditions of life, quite often in cooperation with private subjects. The *Voedseltuin* (Food Garden), Rotterdam, for instance, was supported by Rotterdam's food bank and sponsored by the municipality.

In other cases farmers and gardeners are both able to provide practical support to local administration and to gain, in reward, the possibility to have their say in urban transformation, building, and regeneration processes. Or, again, urban agriculture citywide networks are equally able to establish dedicated agreements for the co-management of vacant green sites, as was the case for *Libere Rape Metropolitane* (Free Metropolitan Beetroots) network in Milan (Silvestri, 2016).

Conclusions

This chapter tentatively suggests that adaptive planning – in its spatial, social, and environmental understanding – can be considered a new approach allowing innovative transcalar and transectorial forms of agency that cross traditional policy boundaries. In facing the need for a new conception of urban governance processes, due to mutable environmental conditions – notably induced by climate change, and the subsequence and increasing pressures over resources, the chapter claims that a revised interpretation of planning processes as a collaborative action involving multiple heterogeneous agents can help the transition towards more sustainable and just practices.

Embedding all actors' needs and values in sustainability planning through a collaborative approach has a major consequence: It is impossible to develop 'one-size-fits-all' or 'silver-bullet' solutions. The solutions always depend on the characteristics of the local communities in crafting sustainable strategies. They are typically place-based and it is crucial to build solutions adapting to the local characteristics.

Nevertheless, these also signal that as the inequality-biased structuring and functioning of social formations (most notably urban deprivation, lack of public decision/engagement, marginalization, etc.) significantly differ in different geographical, economic, and socio-political conditions, innovative solutions that create spaces of justice need to build upon local cultures, needs, and understanding of both sustainability and justice, by prioritizing the goals which, through an adaptive planning process, prove to be most cogent in the context.

In the present chapter brief examples of urban agriculture initiatives have been included to show that an adaptive planning approach is resonant with the need to address the sustainability–justice nexus. These refer to how people take ownership over their own city, how the daily practices of committed citizens can transform neglected spaces into vibrant and inclusive spaces of justice through (sometime temporary) reconceptualization of urban space and power relations. Critical geographers already expanded the point (Hou et al., 2009; Schmelzkopf, 2002; Staeheli et al., 2002), and real-life examples confirm that the power to plan the structure and functioning of the physical space brings along the power to generate spaces of justice/injustice, sustanainability/unsustainability (Bromberg et al., 2007). The diffusion of an adaptive planning approach might imply that the power of socio-ecological (re)production flows through different social agents; and these can weave innovative connections between sustainability and justice. Being ubiquitous and clearly visible in the city space – as literally practiced under the sunlight, urban agriculture initiatives attract a vast interest as an exemplary practice. However, while this chapter specifically refers to urban agriculture with the aim of showing how heterogeneous actors' agency is exerted, the potentiality of adaptive planning has been already pointed out in a number of different areas. For instance, grassroots planning initiatives of temporary indeterminate spaces have been described as able to tackle socio-economic problems in many peripheral areas (Groth & Corijn, 2005). These forms of adaptive planning produced open-access infrastructural solutions that challenge the current mainstream of planning theory (Jiménez, 2014), and present a new framework in which community-driven practice in the making of the public space determines positive outcomes in terms of redistribution, access, and sustainability (Hou & Rios, 2003).

The challenge here is introducing social innovation as a key factor for just transitions to sustainability, so as to include communities of interest, neighbourhood communities and groups of individuals forming voluntary associations, among the main actors of sustainability policies. Transition to sustainability requires social momentum. Promoting collective appropriation of sustainability policies implies that those who will be affected by them are involved in the process of decision-making, right from the beginning and aim at answering the following questions in one way or another: What type of society do they want to live in? Which pathway compromises between the goals and interests of the different groups? Who decides on these compromises?

The Millennium Declaration proclaimed the 'collective responsibility to uphold the principles of human dignity, equality and equity at the global level' (Stokke, 2009). Prior to that, when the United Nations assigned the redaction of a report to the World Commission on Environment and Development in 1983, which is the source of sustainable development, its mission statement mentioned explicitly that its objectives were 'How to reduce inequality and poverty without damaging the environment granted to the future generations' (WCED, 1987). It is now time to go beyond the mantra, and try to do it concretely.

References

Adams, D., Hardman, M., & Larkham, P. (2015). Exploring guerrilla gardening: Gauging public views on the grassroots activity. *Local Environment*, 20(10): 1231–1246.

Agyeman, J., & Erickson, J. (2012). Culture, recognition and the negotiation of difference: Some thoughts on cultural competency in planning education. *Journal of Planning Education and Research*, 32(3): 358–366.

Allen, P. M. (1997). Cities and regions as evolutionary complex systems. *Geographical Systems*, 4: 103–130.

Amin, A., & Thrift, N. (2002). *Cities: Reimagining the Urban*. Cambridge: Polity Press.

Andrews, C. J. (2002). *Humble Analysis: The Practice of Joint Fact-Finding*. Westport, CT: Praeger Publishers.

Batty, M. (2008). Cities as Complex Systems: Scaling, Interactions, Networks, Dynamics and Urban Morphologies. Working Papers Series, University College London.

Beckie, M., & Bogdan, E. (2010). Planting roots: Urban agriculture for senior immigrants. *Journal of Agriculture, Food Systems and Community Development*, 1(2): 77–89.

Been, V., & Voicu, I. (2006). The effect of community gardens on neighboring property values. Law and Economics Working Paper, 46, New York University. http://lsr.nellco.org/nyu_lewp/46

Bettencourt, L. M. A. (2015). Cities as complex systems. In B. A. Furtado, P. A. M. Sakowski, & M. H. Tovolli (eds.), *Modeling Complex Systems for Public Policies*. Brasilia: Institute for Applied Economic Research, 217–236.

Bigot, R. (2009). *Regards sur les quartiers sensibles et les discriminations en France*, Agence nationale pour la cohésion sociale et l'égalité des chances, Collections des rapports, n° 271, CREDOC.

Brenman, M., & Sanchez, T. W. (2012). *Planning as if People Matter: Governing for Social Equity*. Washington, DC: Island Press.

Bromberg, A., Morrow, G., & Pfeiffer, D. (2007). Editorial note: Why spatial justice? *Critical Planning: A Journal of the UCLA Department of Urban Planning/Spatial Justice*, 14: 1–6.

Bulkeley, H., & Betsill, M. (2005). Rethinking sustainable cities: Multilevel governance and the 'urban' politics of climate change. *Environmental Politics*, 14: 42–63.

Certomà, C. (2011). Critical urban gardening as a post-environmentalist practice. *Local Environment*, 16(10): 977–987.

Certomà, C. (2014). Expanding the 'dark side of planning': Governmentality and biopolitics in public garden planning. *Planning Theory*, 14(1): 23–43.

Certomà, C., & Notteboom, B. (2017). Informal planning in a transactive governmentality: Re-reading informal planning practices through Ghent's community gardens. *Planning Theory*, 16/1: 51–73.

Certomà, C., & Tornaghi, C. (2015). Political gardening: Transforming cities and political agency. *Local Environment*, 20/10: 1123–1131.

Christen, M., & Schmidt, S. (2012). A formal framework for conceptions of sustainability: A theoretical contribution to the discourse in sustainable development. *Sustainable Development*, 20(6): 400–410.

Corsín Jiménez, A. (2013). The prototype: More than many and less than one. *Journal of Cultural Economy*, 7/4: 381–398.

COST Action TD1106. (2016). Urban agriculture Europe, EU H2020. www.urban-agriculture-europe.org/

COST Action TU1201. (2016). Urban allotment gardens. Case Studies, EU 2020. www.urbanallotments.eu/case-studies.html

Costa, S., Fox-Kamper, R., Good, R., & Snetic, I. (2016). The position of urban allotment gardens within the urban fabric. In S. Bell, R. Fox-Kämper, N. Keshavarz, et al. (eds.), *Urban Allotment Gardens in Europe*. New York: Abingdon, 201–229.

Crouch, D., & Ward, C. (1997). *The Allotment: Its Landscape and Culture*. Nottingham: Five Leaves.

Daily, G. C., Polasky, S., Goldstein, J., et al. (2009). Ecosystem services in decision making: Time to deliver. *Frontiers of Ecology and Environment*, 7: 21–28.

Donovan, M. G. (2008). Informal cities and the contestation of public space: The Case of Bogotá's street vendors, 1988—2003. *Urban Studies*, 45/1: 29–51

Drilling, M., Giedych, R., & Ponizy, L. (2016). The idea of allotment gardens and the role of spatial and urban planning. In S. Bell, R. Fox-Kämper, N. Keshavarz, et al. (eds.), *Urban Allotment Gardens in Europe*. New York: Abingdon, 35–61.

Eizenberg, E. (2012). Actually existing commons: Three moments of space of community gardens in New York City. *Antipode*, 44(3): 764–782.

Emmett, R. (2011). Community gardens, ghetto pastoral, and environmental justice. *Interdisciplinary Studies in Literature and Environment*, 18(1): 67–86.

FAO (Food and Agriculture Organization of the United Nations). (2013). Urban and peri-urban agriculture. Committee on Agriculture, COAG/99/10. www.fao.org/unfao/bodies/COag/cOAG15/X0076e.htm

Ferris, J., Norman, C., & Sempik, J. (2001). People, land and sustainability: Community gardens and the social dimension of sustainable development. *Social Policy Administration*, 35(5): 559–568.

Fischer, F. (2000). *Citizens, Experts, and the Environment: The Politics of Local Knowledge*. Durham, NC: Duke University Press.

Fischhoff, B., Lichtenstein, S., Slovic, P., Derby, S., & Keeney, R. (1981). *Acceptable Risk*. Cambridge: Cambridge University Press.

Flachs, A. (2010). Food for thought: The social impact of community gardens in the greater Cleveland area. *Electronic Green Journal*, 1(30): 1–9.

Frank, R. (1999). *Luxury Fever: Weighing the Cost of Excess*. Princeton, NJ: Princeton University Press.

Gerometta, J., Haussermann, H., & Longo, G. (2005). Social innovation and civil society in urban governance: Strategies for an inclusive city. *Urban Studies*, 42(11): 2007–2021.

Gibson, R. B. (2006). Sustainability assessment: Basic components of a practical approach. *Impact Assessment and Project Appraisal*, 24: 170–182.

Gibson, R., Hassan, S., Holtz, S., Tansey, J., & Whitelaw, G. (2005). *Sustainability Assessment: Criteria and Processes*. Abingdon, NY: Earthscan Books.

Groth, J., & Corijn, E. (2005). Reclaiming urbanity: Indeterminate spaces, informal actors and urban agenda setting. *Urban Studies*, 42(3): 503–526.

Gueymard, S., & Faburel, G. (2008). *Vécu environnemental et qualité de vie en Région Ile-de-France: une approche exploratoire des inégalités environnementales*. Paris: PUCA.

Harvey, D. (1990). *The Condition of Postmodernity: An Enquiry into the Origins of Cultural Change*. Cambridge, MA: Blackwell Publishing.

Haughton, G., & Hunter, C. (2003). *Sustainable Cities*. New York : Taylor & Francis.

Hou, J. (2010). *Insurgent Public Space: Guerrilla Urbanism and the Remaking of Contemporary Cities*. New York: Taylor & Francis.

Hou, J., Johnson, J., & Lawson, L. (2009). *Greening Cities Growing Communities: Learning from Seattle's Urban Community Gardens*. Seattle: University of Washington Press.

Hou, J., & Rios, M. (2003). Community-driven place making. *Journal of Architectural Education*, 57(1): 19–27.

Howard, E. (1902). *Garden Cities of Tomorrow*. London: S. Sonnenschein & Co. https://archive.org/details/gardencitiestom00howagoog

Humphreys, L. (2010). Mobile social networks and urban public space. *New Media & Society*, 12(5): 763–778.

Ioannou, B., Moran, N., Sondermann, M., Certoma, C., & Hardman, M. (2016). Grassroots movements: Towards cooperative forms of green urban development? In S. Bell, R. Fox-Kämper, N. Keshavarz, et al. *Urban Allotment Gardens in Europe*. New York: Abingdon, 62–90.

IPCC. (2007). Summary for policymakers. www.ipcc.ch/pdf/assessment-report/ . . . /ar4-wg1-spm.pdf

Isgren, E., O'Byrne, D., Jerneck, A., & Ness, B. eds. (2017). Special Issue: Earth System Governance – Task Force Initiative on Sustainability Science. *Challenges in Sustainability*, 5: 1.

Jacobs, J. (1961). *The Death and Life of Great American Cities*. New York: Random House.

Jiménez, A. C. (2014). The right to infrastructure: A prototype for open source urbanism. *Environment and Planning D: Society and Space*, 32(2): 342–362.

Juris, J. S. (2012). Reflections on Occupy Everywhere: Social media, public space, and emerging logics of aggregation. *American Ethnologist*, 39(2): 259–279.

Justus, F., & Taylor, R. (2011). Introduction: Are cities sustainable? In R. Taylor (ed.), *Taking Sides: Clashing Issues in Sustainability*. New York: McGraw-Hill, i–vi.

Klinsky, S., & Golub, A. (2016). Justice and sustainability. In H. Heinrichs, P. Martens, G. Michelsen, & A. Wiek (eds.), *Sustainability Science: An Introduction*. Dordrecht: Springer, 161–173.

Kurtz, H. (2001). Differentiating multiple meanings of garden and community. *Urban Geography*, 22(7): 656–670.

Lamborn, P., & Weinberg, B., eds. (1999). *Avant Gardening: Ecological Struggle in the City and the World*. New York: Autonomedia.

Lang, D. J., Wiek, A., Bergmann, M., et al. (2012). Transdisciplinary research in sustainability science: Practice, principles, and challenges. *Sustainability Science*, 7 (Supplement 1): 25–43.

Larsen, K., Gunnarsson-Östling, U., & Westholm, E. (2011). Environmental scenarios and local-global level of community engagement: Environmental justice, jams, institutions and innovation. *Futures*, 43: 413–423.

Latkowska, M. J., ed. (2015). Scientific Report from Joint Training School 'Social Aspects of Urban Gardening' Warsaw, 7–10 July 2015, COST Action TU 1201, *Urban Allotment Gardens in European Cities* and COST Action TU 1203, *Crime Prevention Through Urban Design and Planning*. www.urbanallotments.eu/training-schools/warsaw-joint-training-school.html

Lefebvre, H. (1991 [1974]). *The Production of Space*. Cambridge, MA: Blackwell Publishing.

Leiserowitz, A., Kates, R., & Parris, T. (2006). Sustainability values, attitudes, and behaviors: A review of multinational and global trends. *Annual Review of Environment and Resources*, 31: 413–444.

Mancebo, F. (2015). Combining sustainability and social justice in the Paris metropolitan region. In C. Isenhour,G. McDonough, & M. Checker (eds.), *Sustainability in the Global City: Myth and Practice*. Series New Directions in Sustainability. Cambridge: Cambridge University Press, 263–283.

Mancebo, F. (2016). Urban agriculture, commons and urban policies: Scaling up local innovation. *Challenges in Sustainability*, 4(1): 10–19.
Mancebo, F., & Sachs, I., eds. (2015). *Transitions to Sustainability*. Dordrecht: Springer.
McKay, G. (2011). *Radical Gardening*. London: Frances Lincoln Limited.
Miller, D. (2002). Debate – Caney's 'International distributive justice': A response. *Political Studies*, 50(5): 974–977.
Miller, J. (2005). Biodiversity conservation and the extinction of experience. *Trends in Ecology & Evolution*, 20(8): 430–434.
Mitchell, D. (2003). *The Right to the City: Social Justice and the Fight for Public Space*. New York: Guilford Press.
Mollinga, P. (2008). Boundary work and the complexity of natural resource management. *Crop Science* 50(Supplement 1): S1–S9.
Moret, J. (2004). La biodiversité à Paris. In Y. Michaud (ed.), *L'université de tous les savoirs*. Paris: Editions Odile Jacob, 217–238.
Moulaert, F., Martinelli, F., González, S., & Swyngedouw, E. (2007). Social innovation and governance in European cities: Urban development between path dependency and radical innovation. *European Urban and Regional Studies*, 14(3): 195–209.
Ostrom, E. (1998). A behavioral approach to the rational choice theory of collective action. *American Political Science Review*, 92(1): 1–22.
Ostrom, E., & Cox, M. (2010). Moving beyond panaceas: A multi-tiered diagnostic approach for social-ecological analysis. *Environmental Conservation*, 37(4): 451–463.
Pagano, M. A., & Bowman, A. (2000). *Vacant Land in Cities: An Urban Resource*. Survey Series. Washington, DC: Brookings Institution.
Pearsall, H., & Pierce, J. (2010). Urban sustainability and environmental justice: Evaluating the linkages in public planning/policy discourse. *Local Environment*, 15: 569–580.
Pudup, M. B. (2008). It takes a garden: Cultivating citizen-subjects in organized garden projects. *Geoforum*, 39(3): 1228–1240.
Quastel, N. (2009). Political ecologies of gentrification. *Urban Geography*, 30(7): 694–725.
Reynolds, K. (2014). Disparity despite diversity: Social injustice in New York City's urban agriculture system. *Antipode*, 47(1): 240–259.
Reynolds, R. (2008). *On Guerrilla Gardening: A Handbook for Gardening without Boundaries*. London: Bloomsbury.
Rittel, H. W., & Webber, M. M. (1973). Dilemmas in a general theory of planning. *Policy Sciences*, 4: 155–169.
Rosol, M. (2012). Community volunteering as neoliberal strategy? Green space production in Berlin. *Antipode*, 44(1): 239–257.
Roy, A. (2005). Urban informality: Toward an epistemology of planning. *Journal of the American Planning Association*, 71(2): 147–158.
Sassen, S. (2006). Making public interventions in today's massive cities. Generalized empowerment: Uneven development and urban interventions. Brussels: City Mine.
Satterthwaite, D. (2007). The transition to a predominantly urban world and its underpinnings. Human Settlements Working Paper Series Urban Change, 4. London: IIED.
Scheffer M., Bascompte J., Brock W. A., et al. (2009). Early-warning signs for critical transitions. *Nature*, 461: 53–59.
Schmelzkopf, K. (2002). Incommensurability, land use, and the right to space: Community gardens in New York City. *Urban Geography*, 23(4): 323–343.
Scholz, R. W. (2011). *Environmental Literacy in Science and Society: From Knowledge to Decisions*. Cambridge: Cambridge University Press.

Seghers, Y., & Van Molle, L. (2007). *Volkstuinen. Een geschiedenis*. Leuven: Davidsfonds.

Silvestri, G. (2016). Community garden networks and their impact on politics and planning – the case of Milan, Italy. In S. Bell, R. Fox-Kämper, N. Keshavarz, et al. (eds.), *Urban Allotment Gardens in Europe*. New York: Abingdon, 70–71.

Smith, C., & Kurtz, H. (2003). Community gardens and politics of scale in New York City. *Geographical Review*, 93(2): 193–212.

Soja, E. W. (1989). *Postmodern Geographies: The Reassertion of Space in Critical Social Theory*. London: Verso.

Soja, E. W. (2000). *Postmetropolis: Critical Studies of Cities and Regions*. Oxford: Basil Blackwell.

Sondermann, M. (2014). *Local cultures of urban gardening and planning in Germany*. In R. Fox-Kämper & N. Keshavarz (eds.), *Riga Event Report: Urban Allotment Gardens in European Cities – Future, Challenges and Lessons Learned (COST Action TU 1201)*, Aachen: Institut für Landes- und Stadtentwicklungsforschung, 15–32.

Staeheli, L., Mitchell, D., & Gibson, K. (2002). Conflicting rights to the city in New York City's community gardens. *GeoJournal*, 58: 197–205.

Stokke, O. (2009). *The UN and Development: From Aid to Cooperation*. UN Intellectual History Project. Bloomington: Indiana University Press.

Swart, R., Robinson, J., & Cohen S. (2003). Climate change and sustainable development: Expanding the options. *Climate Policy*, 3: 19–40.

Tornaghi, C. (2014). Critical geography of urban agriculture. *Progress in Human Geography*, 38: 551–567.

Tracey, D. (2007). *Guerrilla Gardening: A Manualfesto*. Gabriola Island, BC: New Society Publishers.

UN-HABITAT. (2010). State of the world's cities 2010/2011 – Cities for all: Bridging the urban divide. https://unhabitat.org/books/state-of-the-worlds-cities-20102011-cities-for-all-bridging-the-urban-divide/

Van Zeijl-Rozema, A., & Martens, P. (2011). Integrated monitoring of sustainable development. *Sustainability – The Journal of Record*, 4(4): 199–202.

Verweij, M., & Thompson, M. (2007). *Clumsy Solution for a Complex World: Governance, Politics and Plural Perceptions*. Aldershot: Palgrave Macmillan.

Vestbro, D. U. (2013). Development aspect of formal and informal urban types. *Planum: The Journal of Urbanism*, 26(1): 1–12.

Wakefield, S., Yeudall, F., Taron, C., et al. (2007). Growing urban health: Community gardening in south-east Toronto. *Health Promotion International*, 22(2): 92–101.

Walzer, M. (1983). *Spheres of Justice: A Defense of Pluralism and Equality*. New York: Basic Books.

WCED (World Commission on Environment and Development). (1987). *Our Common Future*. Oxford: Oxford University Press.

Weisman, E. L. (2009). Cultivating community: The governance of community gardening in Syracuse, NY. Paper presented at 2009 Joint Meeting of AFHVS and ASFS 2009, College Park, PA, May.

9

Unpacking the Black Box of Urban Climate Agency

(Dis)Empowerment and Inclusion in Local Participatory Processes

SCOTT MORTON NINOMIYA AND SARAH BURCH

New Local Actors Exercising Climate Agency: A Black Box?

Over the past two decades, the growing body of literature on local sustainability transitions indicates that cities have moved from being seen as 'part of the problem' to being key players in the solution (Betsill & Bulkeley, 2007). The focus on cities began to take shape with Agenda 21 at the 1992 Earth Summit and has evolved to encompass the thousands of cities that now have climate action plans (Bulkeley & Betsill, 2005). A milestone in this evolution was the meeting of more than a thousand city mayors at the 21st Conference of the Parties to the United Nations Framework Convention on Climate Change in Paris in 2015 (Ivanova, 2016). A resurgence of local commitment to action also followed President Donald Trump's announcement that the United States would withdraw from the Paris agreement. This resurgence is most strongly evident in thriving local coalitions such as the US Climate Mayors.

This surge of local action is not specific to the United States. In the face of state-level ambivalence and slow progress toward targets, communities around the world are struggling to fill the 'ingenuity gap' between the increasing seriousness of climate change and "the lagging supply of solutions" (Westley et al., 2011). The locus for climate action is thus moving beyond state actors and international treaties (Bäckstrand et al., 2017; Okereke, Bulkeley, & Schroeder, 2013), as a growing number of communities around the world are taking collaborative local action on mitigation, adaptation, and transformation (Bulkeley & Newell, 2015). The literature on this shift to a local locus of climate action is abundant (Bulkeley & Betsill, 2013; Castán Broto & Bulkeley, 2013; Widerberg & Stripple, 2016). The heart of this chapter is the exploration of the many new actors entering the fray to take on new roles, forge new collaborative partnerships, and develop new governance models (Bulkeley & Betsill, 2013; Loorbach & Wijsman, 2013; Newell & Patterson, 2010), with global implications (Hughes, 2016).

This expanding plethora of actors has begun to generate high hopes for participatory processes as vehicles of social innovation for climate action, but the question of whether these high hopes are panning out remains to be answered in current scholarship. Though local agency has expanded rapidly, our understanding of urban agency has been too focused on the municipal state in an unproblematic way, without thinking about the complex networks of agency that have been developed to respond to climate change and how this capacity is being created and harnessed. We suggest that this expanding plethora remains a black box to a certain extent. As local climate governance expands, it is important to begin the crucial task of unpacking the 'what' and 'who' of urban climate agency.

This chapter will begin to delve into the work of unpacking the black box, and provide some concrete ideas on how others can become engaged in that unpacking as well. First, a review of current literature on urban climate governance will outline facets of expanding urban climate agency and identify gaps that point to the need for unpacking. Next, empirical observations from a case study on urban climate governance innovation from Canada to provide starting places for filling the literature gaps give some concrete and practical focus to theoretical discussion by showing real-life examples from emerging local climate action networks. The analysis of the case study will focus on the themes of this book, looking at how local actors are empowered and disempowered, included or excluded in the expanding local networks of the case study. It will also begin the task of investigating how complex networks of local climate agency interact and where there is friction or the potential for friction among the actors in those networks.

Reviewing the Literature: A Multifaceted Black Box to Unpack

This section will outline the exterior angles of the multifaceted black box. State of the art literature is providing an increasingly detailed portrayal of the new actors, processes, networks, and innovations coming onto the local climate scene. Municipal governments are of course among these new actors stepping forward in local transition processes. Scholars have captured the many entry points of municipal governments into the arena of transition, from infrastructure decisions (Creutzig et al., 2016), to adaptation, mitigation, climate action planning, and increasingly more transformative action (Shaw et al., 2014). The past decade has also seen a proliferation of networks that connect municipal level governments around the world. Consequently scholars write extensively on the expansion of such transnational municipal networks (TMNs) (Fünfgeld, 2015; Hayes & Knox-Hayes, 2014), and are increasingly interested in the form and function of local decarbonization initiatives that offer opportunities for more inclusive, imaginative governance of these complex issues.

Local climate action has quickly moved beyond the formal policy channels of municipal governments to include private sector and civil sector actors, as well as scholars (Bulkeley & Betsill, 2013). This local decarbonization space is increasingly polycentric in nature (Meadowcroft, 2007), and navigating this new space presents challenges but also opportunities to expand and reimagine democratic processes to support transformative system entrepreneurship (Jhagroe & Loorbach, 2015). Because these transformative processes are "complex and emergent, and thus beyond traditional control and management" (Burch et al., 2014: 476), no one actor in such a process can steer transformation – contrarily, a plethora of actors exert influence in many different directions.

The proliferation of new actors goes well beyond municipal governments. As Castán-Broto and Bulkeley note: "The literature on global environmental governance now makes clear that non-state actors ... are increasingly involved in responding to climate change" (Castán Broto & Bulkeley, 2013: 93). The desire for tangible, local action has fuelled growing interest as well as financial investment in clean technology solutions and the entrepreneurs who create them (Klewitz & Hansen, 2014; Loorbach & Wijsman, 2013). Actors in the civil sector have also become key players in local climate action (Hawkins & Wang, 2012; Sheppard et al., 2011). Some authors hail the 'third sector' as the key to successful local climate change action (Hale, 2010). Similarly, scholars have become more involved in local climate action, and an expanding cadre of authors urge more academics to become engaged in local decarbonization (Trencher et al., 2014; Wiek & Lang, 2016; Wittmayer & Schapke, 2014). This proliferation of actors, however, adds further layers of complexity to an already complex space. Meadowcroft captures this potential confusion in his work, which asks the question, "Who's in charge here?" and explores some ideas regarding how power can be shared among actors in a radically 'decentred' societal context (Meadowcroft, 2007: 299). Other authors such as Stirling contend that "diversity is rarely an unqualified good. Under any perspective, it presents challenges ... and possible tensions with equity, co-ordination, coherence and accountability" (Stirling, 2011: 86).

This expanding, polycentric array of local actors has also developed a concomitantly broad range of processes to facilitate decarbonization efforts. Local actors are co-creating visions, pathways, and scenarios for decarbonization and setting increasingly ambitious targets for emissions reductions (Hughes, 2016). Visions are stories about desirable future states (Oels, 2009), rather than predictions about likely futures (i.e. forecasts). Even so, quantitative and qualitative data provide crucial input into visioning processes in order to balance desirability with feasibility. These processes, which lead from problem definition to strategy implementation (Wiek & Iwaniec, 2014), include methods such as backcasting (Quist & Vergragt, 2006; Robinson et al., 2011), sustainability solution spaces, and urban living labs (Voytenko et al., 2015).[1]

Visioning processes have been employed around the world to engage actors across a variety of sectors in transition work (McCormick et al., 2014; Talwar et al., 2011). Particularly because transitions take place over decades, the technique of collaboratively generating future scenarios and pathways to realize them can be very helpful to create practical visions for deep decarbonization (Bataille et al., 2016). Examples of participatory scenario building processes abound (Johnson et al., 2012; Schmitt Olabisi et al., 2010), many of which involve backcasting: a popular approach for developing goals or targets for decarbonization and then determining how those goals can be met in within the timeline set out for the target. Participatory backcasting is being used in communities across Europe (Carlsson-Kanyama et al., 2008), and use of this technique is also growing in Canada (Cameron & Potvin, 2016). More recently, some communities have begun to set ambitious targets for deep decarbonization and generating proposed pathways and scenarios to achieve these targets (Burch et al., 2014; Frantzeskaki et al., 2012).

However, there may be limitations to some approaches for developing visions of a decarbonization in the context of a complex, uncertain, and emergent future. Rockström et al. caution that "scenarios often struggle to capture transformative change and the dynamics associated with it: disruption, innovation, and nonlinear change in human behavior" (Rockström et al., 2017: 1269). Others express concern that the inevitable uncertainty of a decarbonized future necessitates prudence about creating blueprints and targets that have many built in, and potentially erroneous, assumptions about how the future will evolve (Cameron & Potvin, 2016; McCormick et al., 2014). Experiments have been offered as a set of strategies that might address this radical uncertainty, by creating spaces within which futures can be envisioned, contested, and enacted (Edwards & Bulkeley, 2017). Castán Broto and Bulkeley present their analysis of 627 urban climate change experiments in a sample of 100 global cities as strong empirical evidence of "experimentation as a key tool to open up new political spaces for governing climate change in the city" (Castán Broto & Bulkeley, 2013: 92). These experiments show promise as a special class of incremental effort with transformative potential (Bulkeley & Castán Broto, 2013).

In addition to processes, communities are utilizing an increasing array of technical, social, and financial innovations to decarbonize energy systems.

[1] For comprehensive examination of key quality criteria that might be applied to sustainability visions, including a review of the scholarship that has contributed to each criterion, see Wiek and Iwaniec (2014).

Adoption of electric vehicles, neighborhood energy systems, and green bonds (to name just a few such developments) allows communities to develop innovative decarbonization solutions at the local level on an unprecedented scale (Avelino, 2011; Bulkeley & Castán Broto, 2013; Widerberg & Stripple, 2016). Despite this flourishing innovative potential, local decarbonization is not a simple matter of fuel switching. Local energy systems are highly carbonized, particularly in Canada, which has among the highest per capita greenhouse gas (GHG) emissions in the world (Burck et al., 2015). As a result, the good life' to which people in North America have become accustomed is inextricably intertwined with these fossil fuel burning energy systems (Burch, 2016). Volumes have been written about the resilience of these systems and the regime resistance that confounds efforts to change them (Geels, 2011, 2014; Meadowcroft, 2011; Schot & Geels, 2008). In this challenging context, local actors who seek to use their expanding agency also need to learn how to navigate the deeply political space of local decarbonization (Burch, 2010). Scholars emphasize that in addition to technical and social innovation, system entrepreneurship is an essential aptitude for local actors who pursue deep decarbonization of local energy systems (Avelino et al., 2013; Olsson et al., 2014). Such system-level innovation involves questioning and challenging the narratives of development that support the powerful status quo of carbonized energy systems (Luederitz et al., 2016). System entrepreneurs must find ways to reduce the resilience of dominant institutions (Westley et al., 2011) and change the conversation about carbon in public discourse at the community level (Cameron & Potvin, 2016; Potvin et al., 2017; Schweizer et al., 2013).

Despite this burgeoning array of literature, systematic learning on local transitions is still early in its development, and O'Brien identifies "the need to develop a critical body of literature on deliberate transformation" (O'Brien, 2012: 667). To build that critical body of literature, scholars need to engage in action research rooted 'on the ground' in communities around the world (Wittmayer & Schapke, 2014). Markard points specifically to a need for more exploration of transformative initiatives in the North American context (Markard et al., 2012).

The expanding field of literature on local climate agency outlined in the foregoing is generally focused on the 'city' as a coherent actor and not delved into the depth of what and who constitutes the 'urban' response and the complexity of generating this urban agency. This gap in the literature has been identified as an area that requires greater attention by several scholars around the world (Bulkeley et al., 2014; Chu et al, 2016; Jordan et al., 2015; Kivimaa et al., 2017; Shi et al., 2016). This chapter therefore contributes to this book and to the Earth System Governance (ESG) Harvesting Initiative by starting to unpack the 'black box' of the urban. The following section will use some of the observations from a local climate initiative in Canada to illustrate how the urban climate change agenda has 'unleashed' new forms of governance arrangement, collaboration, and ways of working at the local level.

To effectively unpack the black box of expanding urban climate agency, scholars must participate in the process of 'learning by doing'. We contend that this approach is crucial to understanding disempowerment and exclusion dynamics in expanding climate agency. Other scholars echo this contention. Wiek et al. (2012) lament that "Sustainability science seems to be still largely 'trapped' in the safe space of descriptive–analytical knowledge production" and call on sustainability scholars to display "a greater willingness to join ... communities to work on practical [sustainability] solutions" (Wiek et al., 2012: 6). Wiek et al. further contend that "Transformational sustainability research foresees a new role for scientists [who] need to immerse themselves into decision processes that are embedded in societal transition processes and build socially robust knowledge" (Wiek et al., 2012: 7). Participatory Action Research (PAR) is an important tool in the toolkit to pursue learning 'on the ground'. Wittmayer and Schapke identify PAR as a new frontier for sustainability research and a promising tool for the collaborative production of knowledge (Wittmayer & Schapke, 2014). A handful of sustainability scholars are exploring the potential of this malleable methodology in sustainability and transitions literature. The next section shows how we brought the PAR approach to bear to glean insights from a Canadian community initiating a decarbonization process.

Unpacking the Black Box of Climate Agency: A Canadian Case Study

Decarbonize Waterloo Region

In November 2016, we (the authors) had the opportunity to be participant observers at the Decarbonize Waterloo Region forum in Waterloo, Ontario. The Decarbonize Waterloo Region forum was designed by another local scholar to bring together diverse stakeholders to discuss and pursue the transformative goal of deep decarbonization of Waterloo Region's energy systems (more details to follow). Through a combination of participant observation and post-forum participant surveys we discovered some intriguing things in that research process and share them here as a contribution to understanding how local actors interact 'inside the black box', with a focus on this book's themes of disempowerment and inclusion. Especially intriguing is our finding that several of the local actors who responded to surveys felt overwhelmed by the magnitude of the task of decarbonization. This points to a strong sense of disempowerment. We also gleaned some significant insights into the theme of inclusion. In this regard, forum participants identified missed opportunities vis-à-vis the absence of key local actors at the forum, and predicted future 'uphill battles'; and potential friction among local actors as the goal of local decarbonization advances into the future. Our goal as PAR scholars was to be constructively critical of the process, and to make a significant contribution not only to the body of scholarly literature, but also to the growing local process of decarbonization. In addition to our learning inside the black box of emerging local climate agency, we also share our initial learning about *how* to unpack the black box, hopeful that it can be applied to advance understanding in many other local contexts and build the ongoing research agenda.

Waterloo Region is a southern Ontario municipality (population approximately 270,000) that has garnered an international reputation for technological and social innovation. The November 2016 forum served as the genesis of the Decarbonize Waterloo Region process. The process was initiated in early 2016 by Heather Douglas, a professor at the University of Waterloo, who sought out a broad range of experts from the local civil sector, academia, all levels of government, energy sector leaders as well as local entrepreneurs. These actors were convened for a two-day invitation-only forum, hosted at the Balsillie School of International Affairs in Waterloo. Douglas articulated ambitious goals for the two-day event: namely, the development of scenarios for fully decarbonized local energy systems and pathways to achieve those scenarios. The scenarios were focused on the three energy-use sectors with the highest GHG emissions, as identified in the Waterloo Region Climate Action Plan – transportation, residential, and industrial/commercial. In addition to the development of scenarios and pathways, the Decarbonize Waterloo Region forum also included the development of policy recommendations for federal, provincial, and municipal governments.

The forum included several presentations by local experts on emerging tools that organizers evaluated as promising for the decarbonization of the Region. These included social innovations like coordinated neighbourhood retrofits for district energy, financial innovations such as impact investment and technical innovations such as geothermal energy systems. A list of policy recommendations for federal, provincial, and local governments was also generated by forum participants in a session designed specifically for that purpose on day two. This discussion included early ideas on how to coordinate policies across the jurisdictions to promote a rapid local transition to decarbonized energy systems. Another session on day two of the forum was dedicated to the articulation of co-benefits that will accrue from efforts to decarbonize the Region's energy systems (in addition to the direct impact of reducing local GHG emissions). The long list included social, economic, and environmental co-benefits.

The paramount goal of the forum was the development of scenarios for local deep decarbonization and pathways to realize those scenarios by 2050. Participants looked at the total energy system of the region quantitatively, using charts prepared before the forum which outlined the energy expended annually in the Region (demand) and the various sources of that energy (supply). Once the groups had developed several potential scenarios in their respective sectors, the next task was to find the 'pathways' to the decarbonized future scenarios. The next section outlines some of the key findings from our research of the forum.

Research Designed to Unpack Local Climate Agency

A participatory action research approach was taken to the study of the Decarbonize Waterloo Region forum, beginning in the planning stages. The research project was designed to engage directly in the transformative efforts of the forum as both a participant and a scholar analyzing the process. The role of the 'immersed' scholar was disclosed and explained to participants by means of a letter in advance of the forum, in addition to in-person introductory remarks at the forum reiterating this role. Throughout the two-day forum, the scholar took part in discussions and took notes about how the process was unfolding.

Immediately after the forum, a 12-question survey was sent electronically to all forum participants. It asked questions about their experience in the forum process: what worked well, what didn't work well, stakeholder interactions and contributions to the process, and what lessons can be learned as the decarbonization process moves forward. Survey participants were recruited by means of an invitation at the forum and a follow up letter. Participation in the survey was optional. Thirty-five out of fifty participants responded to the survey, yielding a 70 per cent response rate. Survey respondents reflected a cross section of the sectors represented at the forum, including civil sector (6), private sector (5), and local government (3). The largest portion of respondents comprised scholars from the two local universities (18). This roughly reflected the same proportion from each sector that was present at the forum, but with an even greater percentage of the overall composition from the academic sector.

Observations from Inside: Local Actors Disempowered

Participant survey responses showed that some participants left the forum feeling overwhelmed by the magnitude of the challenge of deep local decarbonization. Deeper consideration of the bold vision for decarbonization articulated at the forum involved confronting some massive and daunting numbers. Statistics on local energy use and sources compiled for the forum made it clear that Waterloo Region's energy systems are very carbon intensive. More than 77 per cent of combined energy consumed in the residential, transportation, and commercial/industrial sectors comes from burning fossil fuels. In preparation for the forum, statistics on present and future (2050 projections) energy use and sources for the Region were compiled into a series of diagrams and displayed in terms of petajoules. These diagrams provoked a great deal of discussion at the forum (as well as surprise and dismay). Survey responses revealed that many participants found this visualization very helpful for conceptualizing the challenge of decarbonization, but the complexity and concomitant difficulty of the task was noted by many participants. Many participants noted that it was difficult to get their head around the problem – one survey respondent said: "I knew it would be complex – but not this complex." One participant poignantly expressed that this realization left her "a little more grief soaked" than before participating in the forum. This conveys a deep sense of disempowerment.

The forum also uncovered some frustration with the slow pace of progress toward decarbonization to date, and the negative local attitudes that represent a significant barrier to transformative action. In a group discussion early on day one, a local entrepreneur denounced stifling local attitudes that he had encountered towards decarbonization, proclaiming, "I'm sick and tired of being told it can't be done here. I see so many things happening in other places."

Several other participants also expressed concern with the progress to date toward the Plan's incremental goal of 6 per cent reductions in greenhouse gas emissions by 2020 in the region. One survey respondent said it was, "great that our [policy] recommendations are in alignment with a document that has been approved by all 3 municipalities and the region [Climate Action Plan], but not so great that things seem to be in the same place as 3 years ago." Another respondent was more critical, warning that, "If we are content to leave this in the hands of policy makers without the development of concrete steps that stakeholders need to take, we will likely be in the same situation in another 3 years" [referring to the three years between the adoption of the current Climate Action Plan in 2013 and the forum in 2016]. The bold vision for 100 per cent decarbonization by 2050 was described as inspiring by many participants but many also expressed consternation about the complexity and scale of achieving that bold vision. Though the current local approach of an incremental 6 per cent reduction target by 2020 was criticized as inadequate, participants acknowledged that the task of discerning concrete next steps in a more transformative approach felt unfinished at the end of the forum.

Observations from Inside: Local Actors Excluded

Exclusion/inclusion was also a theme that emerged strongly in our findings. Several participants recognized the reality that the Decarbonize Waterloo Region forum did not engage a full representation of Waterloo Region's population. One survey respondent even labelled himself and his fellow forum participants as "an elitist cohort." Other members of that cohort responded in the survey that key actors were missing: people from rural areas, those without post-secondary education. Other survey respondents also noted as missing, "ordinary citizens who will be affected by changes – especially those on the lower income spectrum." Another survey respondent further problematized this dynamic, claiming "it would have been good to have more dissenting views in the room." Another respondent lamented that the forum, "lacked naysayers to challenge us further." While many participants lauded the collaborative space that was fostered at the forum, there was also strong concern expressed that there must be ongoing efforts to engage dissenting voices who might contest the radical changes required for deep decarbonization. Observations like these point to a recognition of the reality of future friction between the interests and needs of local actors and a need to navigate that friction carefully.

The fact that no elected leaders attended was noted with dismay by some, who lamented the lack of "people with decision power," and warned that it was "all well for those in the room to say what should be done, but no one with the ability to make real change was there." Interestingly, several respondents expressed relief at the "lack of political or ideological agenda" at the forum, but others noted that without the incumbents on board, it will be very difficult to move innovative solutions forward. Participant survey responses revealed that there is an understanding of the importance of broader inclusion in the process, including local energy incumbents such as the electrical utilities. Ten participants expressed disappointment with the low representation from energy incumbents such as electrical and gas utilities and energy-intensive industries.

Participants demonstrated a keen awareness that more work must also be done to determine how to engage the broader public in the vision and implementation of decarbonization. In the surveys, a respondent expressed concerns about gaining political and public support for decarbonization, saying "it's still a matter of convincing local folks that we really need to do this!" Still another respondent asserted that the forum should have included more "discussion of how to build popular support that endures the political winds of change." Along the same theme, another section of the forum agenda focused specifically on the political aspects of decarbonization, generating ideas for policy change at the local, provincial and federal levels. Participants applauded this work, but also cautioned that, "While good energy was put into the list of desired policies, there wasn't discussion around the fact that without developing and demonstrating popular support for the policies, policy makers will have difficulty implementing them." One respondent had some interesting ideas about how to engage the public and build that popular support. That respondent reported that the forum "showed me that the focus may not need to be on technologies so much as culture change. I am wondering what that forum might look like. Perhaps after seeing what comes out of the [Waterloo Region] Community Energy Investment Strategy, if some decisions are made locally about what needs to be done, a forum about how to bring our community along with it would be a good thing to do."

Observations from Inside: Openings for Empowerment and Inclusion

The evolution from an exclusive 'elitist cohort' feeling disempowered to a more inclusive and empowered collective of local actors is the next grand task for the Canadian community of Waterloo Region. We must note that the picture that emerged from our research at the forum is not entirely bleak. Survey responses reflected that many participants shared the sentiment that the November 2016 forum set the stage for ongoing constructive dialogue. Participants expressed excitement and hope about the potential for collaboration among the diverse experts around the vision for local decarbonization. Seven survey respondents articulated a strong appreciation for the "excellent mix of people, mix of disciplines" and the growing potential for polycentric leadership on decarbonization in Waterloo Region. Several survey respondents lauded the collaborative and respectful discussion, with one asserting that the "tone of conversation was very supportive of everyone."

The forum also provided a platform for sharing learning and practical expertise on energy system innovations that could be helpful for moving decarbonization forward. These included presentations on geothermal energy, hydrogen vehicles, neighbourhood heating systems, and other innovative ideas. These presentations were regarded as very useful and enlightening, as summarized by one respondent who appreciated that the "technical presentations provided a toolbox of possible solutions for us to draw on, and showcased efforts specific to the region." Many other respondents expressed interest and excitement about the growing array of innovative technological solutions emerging at the local level.

Finally, the forum also helped to engender a strong sense of inspiration to take bold local action.

Despite the widespread sentiments of feeling overwhelmed, participants expressed a sense of hope and desire for ongoing engagement in local decarbonization. Surveys showed clearly that participants were willing to stay involved in a variety of ways. In addition, 100 per cent of survey respondents expressed willingness to take part in follow-up focus groups to explore next steps in the decarbonization process. Eight survey respondents expressed a desire to ramp up local efforts significantly saying that the forum reminded them "that there are a multitude of systems that need to fundamentally shift. It showed the importance of acting now to lay a foundation for future, deeper change." Another participant shared in a forum plenary session that "people go where systems led them – we have to change those systems." Another survey respondent expressed appreciation for the panel presenter (an academic) who "inspired us to be bold." This respondent also conveyed that the forum was "a reminder that we need to stop burning fossil fuels – i.e. just making something a bit more efficient or cleaner isn't going to get us where we need to be." One respondent made a particularly powerful statement about the vision for deep decarbonization: "I have to say how thrilled I am that we had a forum called Decarbonizing Waterloo Region – I think that in itself is an incredibly important step forward in articulating what we need to do." The clear and complex challenge for the community now is to convert the inspiration engendered at the forum into empowerment, inclusion and – above all – action for deep decarbonization.

Conclusion

Harvesting Learning, Sowing More Questions

This look inside the black box of local climate agency yielded insights into disempowerment, exclusion, and potential for friction among actors in one local community. Some openings for empowerment and inclusion were also revealed. These insights help to fill the gap identified in the literature review by endeavouring to unpack the black box of climate agency in one community. Analysis of these insights gives rise to questions that researchers can put forth in future research into the 'black box' in communities around the world.

There is more participatory action research to do in Waterloo Region on the roles that each of these actors can play in mobilizing transformative potential at the local level. Local scholars have shown potential to act as provocateurs of discussion and action on more transformative goals like deep decarbonization. Entrepreneurs may also show emerging promise as passionate advocates of innovation. The local civil sector has already shown its potential to bring many actors together and achieve consensus on the Climate Action Plan. Local government was also recognized at the forum as key leaders. This emerging polycentric approach also provides an opportunity to learn how to make the most out of each actor's strengths. This gives rise to an important question for ongoing research in other communities: How can the multiple local actors listed here be included and empowered effectively to make the most of each of these emerging strengths?

There will undoubtedly be winners and losers in the process of deep decarbonization and transforming energy systems will have massive social and economic impacts that must be carefully investigated throughout the process. Questions of justice and equity must also be core considerations. For that reason, it is crucial to invest energy in the expansion of local decarbonization dialogue and action beyond the "elitist cohort" that gathered at the November 2016 forum. There are many aspects of diversity that also must be considered to make local decarbonization a more democratic process of co-production, including education level, socio-economic status, culture, urban/rural, among others. Waterloo Region is a very diverse community and the full picture of that diversity should be reflected in the decarbonization process. This gives rise to a second important question for ongoing research in other communities: How can local actors build more inclusive processes?

Finally, as the emerging collective driving Decarbonize Waterloo Region seeks empowerment and inclusion, it must carefully navigate the space held by the 'agency incumbents'. These incumbents – the utilities, elected officials and the public – who generate energy, policy, and public discourse in Waterloo Region, hold a great deal of power (Avelino & Wittmayer, 2016; Geels et al., 2014). This gives rise to a third important question for ongoing research in other communities: How can local actors engage energy and power incumbents effectively while maintaining a radical trajectory of deep decarbonization?

Going forward, the outcomes of this decarbonization initiative are uncertain. Finding answers to the questions highlighted in this section will be crucial to future success, both locally, and in other communities where local actors seek to expand their climate agency. In this era when the local level is the ascendant locus of climate action, the lessons learned in Waterloo Region have global implications and applications.

In addition to the insights elaborated thus far in this section, this research process also provided insights into the 'how' of unpacking the black box of emerging local climate agency. In other words, there is much to learn from the process of learning that was undertaken here in this nascent effort. Taking a reflexive stance, we can identify both promising, problematic approaches in the research that we undertook. These will be discussed in the following section, and included in our suggestions for building that ongoing research agenda.

Future Research Agenda: Assembling a Toolkit to Unpack the Black Box

Participatory, multiactor processes have sprung up around the world, and the Decarbonize Waterloo Region process presented an interesting opportunity for learning about their inner workings. The Decarbonize Waterloo Region forum process was an imperfect opening – but an opening nonetheless. The insights gleaned in our research can help to inform the development of an ongoing research agenda to be pursued in communities around the world for decades to come.

Empowerment: Developing a Toolkit for Constructive Critique. An important item on that agenda is the cultivation of develop deeper skills in the PAR approach as researchers – a toolkit for unpacking the black box. Wittmayer et al. advise that PAR approaches "blur traditional role understandings and raise questions with regard to training requirements (i.e. Which competencies are needed?), quality criteria (i.e. What are appropriate quality standards for this kind of research?) and intervention legitimacy (i.e. What kind of intervention is legitimate by whom and why?)" (Wittmayer & Schapke, 2014: 484).

Also, as Wiek et al. urge, scholars need to get involved in local climate action and learn more nuanced methods to critique constructively in ways that help to empower local actors to build solutions. It is not helpful or ethical to simply observe and critique dispassionately, pronouncing 'success' or 'failure' in long, complex, and non-linear processes of experimentation and learning-by-doing that foster local transformations. To do so would constitute an ultimate act of disempowerment. There will likely be friction between a more traditional academic stance of dispassionate critique and an approach of scholarship that engages directly with other local actors to drive local climate action processes. Scholars will need to develop strategies to navigate that potential friction.

Inclusion: Sharing Tools, Building Capacity and Engaging Diverse Actors in Research. Another important research agenda item is the matter of inclusion and capacity development. Robinson et al. insist that "for a truly consultative and consensus-oriented process to occur, it is important that a broad sample of the community be engaged in the discussion that are equipped with technical knowledge or understanding of the goals of the process in order to participate in an equitable and effective fashion" (Robinson et al., 2011: 766). This imperative to expand participation opens yet another opportunity to employ a PAR approach. As MacDonald reflects, "the purpose of PAR is to foster capacity, community development, empowerment, access, social justice, and participation" (MacDonald, 2012: 48). To foster capacity, many more scholars need to be encouraged to equip themselves with the tools that they need to unpack the black box discussed in this chapter. Further, much more needs to be learned about how to use PAR to engage an increasingly diverse cohort in the work of unpacking. Scholars are not the only ones who should be encouraged or 'allowed' to wield the tools of research. All the actors discussed in this chapter need to be engaged in this work, including entrepreneurs, government and elected representatives, civil sector actors, and others. Engaging and incorporating different 'ways of knowing' into the ongoing research agenda can help to uncover deep and poignant insights that inform the kind of innovative approaches that are so desperately required.

Final Words

Moving into an uncertain future the local actors in Waterloo Region and other communities around the world need to create conditions that can facilitate deep decarbonization. They can do so by moving forward boldly to harness this potential, prudently to overcome the barriers, and reflexively to respond to the complex and uncertain future as it unfolds. This chapter analyzed observation and interaction with the local actors at the Decarbonize Waterloo Region forum to produce insights on exclusion and disempowerment in a deep decarbonization initiative in a Canadian community. As such, the insights from the early work toward decarbonization in Waterloo Region align with, and build upon, the growing and diverse domain of ESG scholarship. We have built upon more established insights into the multilevel governance of climate change in urban spaces, the equity (or allocation) implications of various low-carbon transition pathways, and the governance architectures that might foster inclusive and flexible responses to complex sustainability challenges. Future work in this domain will build on these insights to explore strategies for building inclusive participatory processes that more effectively generate imaginative outcomes, which are then tied to specific tools and strategies. Ultimately, they will contribute to a deeper understanding of the inner working of local climate agency so that this rapidly proliferating field can make maximum global impact at this crucial historical juncture.

References

Avelino, F. (2011). Power in transition: Empowering discourses on sustainability. *Transitions* (January): 405.

Avelino, F., & Wittmayer, J. M. (2016). Shifting power relations in sustainability transitions: A multi-actor perspective. *Journal of Environmental Policy & Planning*, 18(5): 628–649.

Avelino, F., Wittmayer, J., Haxeltine, A., et al. (2013). Game changers and transformative social innovation: The case of the economic crisis and the new economy. TRANSIT working paper (613169): 1–24.

Bäckstrand, K., Kuyper, J. W., Linnér, B., et al. (2017). Non-state actors in global climate governance: From Copenhagen to Paris and beyond. *Environmental Politics*, 26(4): 561–579. http://doi.org/10.1080/09644016.2017.1327485

Bataille, C., Waisman, H., Colombier, M., Segafredo, L., Williams, J., & Jotzo, F. (2016). The need for national deep decarbonization pathways for effective climate policy. *Climate Policy*, 16(Supplement 1): S7–S26.

Betsill, M., & Bulkeley, H. (2007). Looking back and thinking ahead: A decade of cities and climate change research. *Local Environment*, 12(5): 447–456.

Biermann, F., Abbott, K., & Andresen, S. (2012). Navigating the Anthropocene: Improving earth system governance. *Science*, 335(6074): 1306–1307. http://doi.org/10.1126/science.1217255

Bulkeley, H., & Betsill, M. (2005). Rethinking sustainable cities: Multilevel governance and the 'urban' politics of climate change. *Environmental Politics*, 14(1): 42–63.

Bulkeley, H., & Betsill, M. M. (2013). Revisiting the urban politics of climate change. *Environmental Politics*, 22(1): 136–154. http://doi.org/10.1080/09644016.2013.755797

Bulkeley, H., & Castán Broto, V. (2013). Government by experiment? Global cities and the governing of climate change. *Transactions of the Institute of British Geographers*, 38 (3): 361–375. http://doi.org/10.1111/j.1475–5661.2012.00535.x

Bulkeley, H., Edwards, G. A., & Fuller, S. (2014). Contesting climate justice in the city: Examining politics and practice in urban climate change experiments. *Global Environmental Change*, 25: 31–40.

Bulkeley, H., & Newell, P. (2015). *Governing Climate Change*. London: Routledge.

Burch, S. (2010). Transforming barriers into enablers of action on climate change: Insights from three municipal case studies in British Columbia, Canada. *Global Environmental Change*, 20(2): 287–297. http://doi.org/10.1016/j.gloenvcha.2009.11.009

Burch, S. (2016). *Will Ontario's Climate Change Action Plan Transform Communities?* CIGI Policy Brief No. 90. Waterloo: Centre for International Governance Intervention.

Burch, S., Shaw, A., Dale, A., & Robinson, J. (2014). Triggering transformative change: A development path approach to climate change response in communities. *Climate Policy*, 14(4): 467–487. http://doi.org/10.1080/14693062.2014.876342

Burck, J., Marten, F., Bals, C., et al. (2015). The Climate Change Performance Index Results. https://germanwatch.org/en/download/13626.pdf%5Cnhttps://germanwatch.org/en/ccpi

Cameron, L., & Potvin, C. (2016). Characterizing desired futures of Canadian communities. *Futures*, 82: 37–51. http://doi.org/10.1016/j.futures.2016.05.003

Carlsson-Kanyama, A., Dreborg, K. H., Moll, H. C., & Padovan, D. (2008). Participative backcasting: A tool for involving stakeholders in local sustainability planning. *Futures*, 40(1): 34–46.

Castán Broto, V., & Bulkeley, H. (2013). A survey of urban climate change experiments in 100 cities. *Global Environmental Change*, 23(1): 92–102. http://doi.org/10.1016/j.gloenvcha.2012.07.005

Chu, E., Anguelovski, I., & Carmin, J. (2016). Inclusive approaches to urban climate adaptation planning and implementation in the Global South. *Climate Policy*, 16(3): 372–392.

Creutzig, F., Agoston, P., Minx, J. C., et al. (2016). Urban infrastructure choices structure climate solutions. *Nature Climate Change*, 6(12): 1054. http://doi.org/10.1038/nclimate3169

Edwards, G., & Bulkeley, H. (2017). Heterotopia and the urban politics of climate change experimentation. *Environment and Planning D: Society and Space*, 36(2): 350–369. http://dro.dur.ac.uk/23804/

Frantzeskaki, N., Loorbach, D., & Meadowcroft, J. (2012). Governing societal transitions to sustainability. *International Journal of Sustainable Development*, 15(1–2): 19–36.

Fünfgeld, H. (2015). Facilitating local climate change adaptation through transnational municipal networks. *Current Opinion in Environmental Sustainability*, 12: 67–73.

Geels, F. W. (2011). The multi-level perspective on sustainability transitions: Responses to seven criticisms. *Environmental Innovation and Societal Transitions*, 1(1): 24–40. http://doi.org/10.1016/j.eist.2011.02.002

Geels, F. W. (2014). Regime resistance against low-carbon transitions: Introducing politics and power into the multi-level perspective. *Theory, Culture & Society* (May 2013). http://doi.org/10.1177/0263276414531627

Hale, S. (2010). The new politics of climate change: Why we are failing and how we will succeed. *Environmental Politics*, 19(2): 255–275.

Hawkins, C. V., & Wang, X. (2012). Sustainable development governance: Citizen participation and support networks in local sustainability initiatives. *Public Works Management & Policy*, 17(1): 7–29.

Hayes, J., & Knox-Hayes, J. (2014). Security in climate change discourse: Analyzing the divergence between US and EU approaches to policy. *Global Environmental Politics*, 14(2): 82–101.

Hughes, S. (2016). The politics of urban climate change policy: Toward a research agenda. *Urban Affairs Review*, 1–19. http://doi.org/10.1177/1078087416649756

Ivanova, M. (2016). Good COP, bad COP: Climate reality after Paris. *Global Policy*, 7(3): 411–419.

Jhagroe, S., & Loorbach, D. (2015). See no evil, hear no evil: The democratic potential of transition management. *Environmental Innovation and Societal Transitions*, 15: 65–83. http://doi.org/10.1016/j.eist.2014.07.001

Johnson, K. A., Dana, G., Jordan, N. R., et al. (2012). Using participatory scenarios to stimulate social learning for collaborative sustainable development. *Ecology and Society*, 17(2): 9.

Jordan, A. J., Huitema, D., Hildén, M., et al. (2015). Emergence of polycentric climate governance and its future prospects. *Nature Climate Change*, 5(11): 977.

Kivimaa, P., Hildén, M., Huitema, D., Jordan, A., & Newig, J. (2017). Experiments in climate governance–a systematic review of research on energy and built environment transitions. *Journal of Cleaner Production*, 169: 17–29.

Klewitz, J., & Hansen, E. G. (2014). Sustainability-oriented innovation of SMEs: A systematic review. *Journal of Cleaner Production*, 65: 57–75.

Loorbach, D., & Wijsman, K. (2013). Business transition management: Exploring a new role for business in sustainability transitions. *Journal of Cleaner Production*, 45: 20–28.

Luederitz, C., Schäpke, N., Wiek, A., et al. (2016). Learning through evaluation – A tentative evaluative scheme for sustainability transition experiments. *Journal of Cleaner Production*. http://doi.org/10.1016/j.jclepro.2016.09.005

MacDonald, C. (2012). Understanding participatory action research: A qualitative research methodology option. *Canadian Journal of Action Research*, 13(2): 34–50. http://ezproxy.lib.ucalgary.ca/login?url=http://search.ebscohost.com/login.aspx?direct=true&db=ehh&AN=89233336&site=ehost-live

Markard, J., Raven, R., & Truffer, B. (2012). Sustainability transitions: An emerging field of research and its prospects. *Research Policy*, 41(6): 955–967. http://doi.org/10.1016/j.respol.2012.02.013

McCormick, K., Gaziulusoy, I., Twomey, P., & McGrail, S. (2014). Decarbonization of cities: You're dreaming! (December), 12–15.

Meadowcroft, J. (2007). Who is in charge here? Governance for sustainable development in a complex world. *Journal of Environmental Policy & Planning*, 9(3–4): 299–314. http://doi.org/10.1080/15239080701631544

Meadowcroft, J. (2011). Engaging with the politics of sustainability transitions. *Environmental Innovation and Societal Transitions*, 1(1): 70–75. http://doi.org/10.1016/j.eist.2011.02.003

Newell, P., & Paterson, M. (2010). *Climate Capitalism: Global Warming and the Transformation of the Global Economy*. Cambridge: Cambridge University Press.

O'Brien, K. (2012). Global environmental change II: From adaptation to deliberate transformation. *Progress in Human Geography*, 36(5): 667–676. http://doi.org/10.1177/0309132511425767

Oels, A. (2009) The power of visioning: The contribution of future search conference to decision-making in Local Agenda 21 processes. In F. Coenen (ed.), *Public Participation and Better Environmental Decisions*. New York: Springer Science+Business Media, 73–88.

Okereke, C., Bulkeley, H., & Schroeder, H. (2013). Conceptualizing climate governance beyond the international regime. *Global Environmental Politics*, 9(1): 58–78.

Olsson, P., Galaz, V., & Boonstra, W. J. (2014). Sustainability transformations: A resilience perspective. *Ecology and Society*, 19(4). http://doi.org/10.5751/ES-06799–190401

Pelling, M., O'Brien, K., Matyas, D. (2015) Adaptation and transformation. *Climatic Change*, 133: 113–127.

Potvin, C., Sharma, D., Creed, I., et al. (2017). Stimulating a Canadian narrative for climate. *FACETS*, 2: 131. http://doi.org/10.1139/facets-2016–0029

Quist, J., & Vergragt, P. (2006) Past and future of backcasting: The shift to stakeholder participation and a proposal for a methodological framework. *Futures*, 38: 1027–1045.

Robinson, J., Burch, S., Talwar, S., O'Shea, M., & Walsh, M. (2011). Envisioning sustainability: Recent progress in the use of participatory backcasting approaches for sustainability research. *Technological Forecasting and Social Change*, 78(5): 756–768. http://doi.org/10.1016/j.techfore.2010.12.006

Rockström, J., Gaffney, O., Rogelj, J., Meinshausen, M., Nakicenovic, N., & Schellnhuber, H. J. (2017). A roadmap for rapid decarbonization. *Science*, 355 (6331): 1269–1271.

Rockström, J., Steffen, W., Noone, K., et al. (2009). A safe operating space for humanity. *Nature*, 461 (September): 472–475.

Schmitt Olabisi, L. K., Kapuscinski, A. R., Johnson, K. A., Reich, P. B., Stenquist, B., & Draeger, K. J. (2010). Using scenario visioning and participatory system dynamics modeling to investigate the future: Lessons from Minnesota 2050. *Sustainability*, 2(8): 2686–2706.

Schot, J., & Geels, F. W. (2008). Strategic niche management and sustainable innovation journeys: Theory, findings, research agenda, and policy. *Technology Analysis & Strategic Management*, 20(5): 537–554. http://doi.org/10.1080/09537320802292651

Schweizer, S., Davis, S., & Thompson, J. L. (2013). Changing the conversation about climate change: A theoretical framework for place-based climate change engagement. *Environmental Communication: A Journal of Nature and Culture*, 7(1): 42–62. http://doi.org/10.1080/17524032.2012.753634

Shaw, A., Burch, S., Kristensen, F., Robinson, J., & Dale, A. (2014) Accelerating the sustainability transition: Exploring synergies between adaptation and mitigation in British Columbian communities. *Global Environmental Change*, 25: 41–51.

Sheppard, S. R., Shaw, A., Flanders, D., et al. (2011). Future visioning of local climate change: A framework for community engagement and planning with scenarios and visualisation. *Futures*, 43(4): 400–412.

Shi, L., Chu, E., Anguelovski, I., et al. (2016). Roadmap towards justice in urban climate adaptation research. *Nature Climate Change*, 6(2): 131.

Steffen, W., Richardson, K., Rockström, J., et al. (2015). Planetary boundaries: Guiding human development on a changing planet. *Science* (New York), 348(6240): 1217. http://doi.org/10.1126/science.aaa9629

Stirling, A. (2011). Pluralising progress: From integrative transitions to transformative diversity. *Environmental Innovation and Societal Transitions*, 1(1): 82–88.

Trencher, G., Bai, X., Evans, J., McCormick, K., & Yarime, M. (2014). University partnerships for co-designing and co-producing urban sustainability. *Global Environmental Change*, 28: 153–165.

Voytenko, Y., McCormick, K., Evans, J., & Schliwa, G. (2015). Urban living labs for sustainability and low carbon cities in Europe: Towards a research agenda. *Journal of Cleaner Production*, 123: 45–54.

Westley, F., Olsson, P., Folke, C., et al. (2011). Tipping toward sustainability: Emerging pathways of transformation. *Ambio*, 40(7): 762–780. http://doi.org/10.1007/s13280-011–0186-9

Widerberg, O., & Stripple, J. (2016). The expanding field of cooperative initiatives for decarbonization: A review of five databases. *Wiley Interdisciplinary Reviews: Climate Change*, 7(4): 486–500. http://doi.org/10.1002/wcc.396

Wiek, A., & Iwaniec, D. (2014) Quality criteria for vision and visioning in sustainability science. *Sustainability Science*, 9: 497–512.

Wiek, A., Ness, B., Schweizer-Ries, P., Brand, F. S., & Farioli, F. (2012). From complex systems analysis to transformational change: A comparative appraisal of sustainability science projects. *Sustainability Science*, 7(Supplement 1): 5–24. http://doi.org/10.1007/s11625-011–0148-y

Wittmayer, J. M., & Schäpke, N. (2014). Action, research and participation: Roles of researchers in sustainability transitions. *Sustainability Science*, 9(4): 483–496. http://doi.org/10.1007/s11625-014-0258-4

10

From Public to Citizen Responsibilities in Urban Climate Adaptation

A Thick Analysis

CAROLINE J. UITTENBROEK, HELEEN L. P. MEES, DRIES L. T. HEGGER, AND PETER P. J. DRIESSEN

Introduction

Climate change is a reality. Insights on climate change impacts worldwide show that climate change will pose increasing challenges to cities (Adger et al., 2003; Rockström et al., 2009). Even in the unlikely event of a short-term fundamental alteration of current production and consumption practices and decreasing carbon dioxide (CO_2) emissions, there will be a need to cope with extreme weather events that will become more frequent (Runhaar et al., 2012). These events include extreme storms, fluvial and pluvial floods, stronger urban heat waves, and longer dry periods, among other climate impacts. Increasing urbanization will exacerbate the potential consequences of such events (Hegger et al., 2014; Runhaar et al., 2012).

Until the beginning of the twenty-first century, debates in the literature and practice focused predominantly on the need to mitigate climate change. But from then on, a vast literature emerged on the need to accompany mitigation efforts with climate change adaptation (Adger et al., 2009; Mees, 2017). This literature has addressed several different issues regarding the governance of climate change adaptation. To contextualize this literature, it is important to note the differences between world regions in how adaptation governance is discussed, both in academic literature and practice. Literature on adaptation in the Global South focuses on adaptation by and within communities (community-based adaptation) (Olsson et al., 2015). Literature on the Global North focuses more on adaptation policy and actions by governments, households, and individual citizens. Within literature focusing on the Global North, there is an important distinction in the roles attributed to governmental actors vis-à-vis other actors. In North America and predominantly the USA the default position is that adaptation should be pursued by individuals and companies, in short, that it is private actors' responsibility. In Western Europe, with its long welfare state legacy, the assumption in debates in literature and practice is that governmental actors have an important duty of care (Driessen & van Rijswick, 2011). These two distinctions are important to keep in mind to be able to interpret debates on adaptation governance. The current chapter draws on literature and practice with a focus on the Global North and discusses literature and empirical examples from the Netherlands, a prosperous Western European country.

In this context, among other issues, the literature has addressed questions on the kinds of approaches that should be used. For instance, is a dedicated approach desirable, in which predominantly governmental actors develop and implement specific 'adaptation policies' (Jordan & Lenschow, 2010)? Or is a 'mainstreaming' approach preferable, in which climate change adaptation is routinely embedded in other policy domains (Uittenbroek et al., 2013)? The adaptation governance literature has also focused on the issue of agency, i.e. the role of different public and private actors in adaptation (Mees, 2017). It was shown that in practice it is often local governments that take the lead in making concrete local policies or taking specific actions (Hegger et al., 2017; Mees, 2017).

With respect to agency, a recent strand in the debate on the governance of climate change adaptation is that of citizen responsibilization (Mees et al., 2012; Tompkins & Eakin, 2012). Responsibilization of citizens is often associated with "... how politicians and governments publicly frame and legitimize a new realm of state intervention dedicated to enticing, persuading and nudging citizens to 'take responsibility' in producing public value" (Peeters, 2013: 586). Citizen responsibilization signals towards a shift from welfare state collectivism to the responsibilization of individuals (Ilcan & Basok, 2004), and is criticized by many for being a hyper-individualist and depoliticized brand of neoliberal governmentality (Kistner, 2009). In the context of this debate, it is increasingly argued that citizens need to take more responsibilities in climate change adaptation to decrease the burden on local governments (Hegger et al., 2017; Wamsler & Brink, 2014). A substantive argument put forward in favour of citizen responsibilization is that governmental actors do not have the capacity or the authority to act on private properties, while many adaptation measures such as urban green or decoupling of rainwater from the sewage system by reducing the amount of hardened surface can and should be taken on such private properties (Tompkins & Eakin, 2012). Besides that, also more normative arguments for citizen responsibilization are put forward. One of these is the position that governmental actors should limit themselves to those actions that cannot be taken by individuals or groups of citizens (Driessen & van Rijswick, 2011). Arguably, this argument is sometimes made in the context of a broader neo-liberal political agenda and budget cuts, as mentioned before. Another argument is that citizen responsibilization may increase community resilience (Driessen & van Rijswick, 2011).

It is often implicitly suggested that such citizen responsibilization will lead to citizen empowerment and that it will enable societies to make better use of the 'energy' that is present in what Hajer (2011) has aptly termed 'the energetic society'. These are assumptions that deserve to be problematized and critically reviewed. Among other concerns, there is the risk that citizen responsibilization might lead to inequalities in the sense that some (arguably well-off) citizens become empowered while others become disempowered (Driessen & van Rijswick, 2011). There is also the risk of maladaptation (i.e. that the results of an action, directly or indirectly, increase the vulnerability to climate change rather than reduce as is intended with adaptation) (Wamsler & Brink, 2014) and citizen fatigue as a result of overcharging citizens with responsibilities (for which they might not have the necessary time, knowledge, etc.).

This chapter will address these concerns by providing a critical review of current academic debates about and practice of citizen responsibilization in urban climate change adaptation. To achieve the chapter's goal, it will take the following steps. First, it provides a brief state of the art of insights from the domain of environmental governance both on citizen responsibilization more generally and on urban climate change adaptation more specifically. Next, we introduce a prominent framework for evaluating the quality of environmental decision making – Adger et al.'s (2003) thick analysis that incorporates the criteria of economic efficiency, environmental effectiveness, equity, and political legitimacy. The chapter goes on by providing an illustrative practical example of citizen responsibilization in climate adaptation, namely Dutch experiences. The examples illustrate the tensions regarding citizen responsibilization that have been documented in the literature and also shows the potential synergies and trade-offs between the four evaluation criteria. We conclude this chapter by briefly summarizing the current state of the art in literature and providing an overview of approaches that are currently proposed to deal with the observed tensions.

The Shift from Public to Private Responsibilities in Climate Adaptation: A State of the Art

With the rise of the neo-liberal agenda and its wave of privatization in the 1980s the automatic link between public issues and the public domain was challenged (Mees, 2017). It is now increasingly accepted that the responsibility for public issues can be shared with or even completely transferred to private actors (Dubbink, 2003). The shift from public to private responsibilities for public issues is a main feature of the 'shift from government to governance' as discussed in the governance literature (e.g. Jordan et al., 2005; Rhodes, 2007; van Kersbergen & van Waarden, 2001). Power and authority are transferred not only from the government downward and upward to other levels of government, but also outward to private actors. New governance arrangements have emerged in which governments have a 'steering' rather than 'rowing' role (Mees, 2017). Nevertheless, governance scholars also claim that the concept of governance is vague, and often it is more an ideal or normative prescription than that it reflects empirical reality (Arts, 2014; Capano et al., 2015; Jordan et al., 2005). While there are numerous and varying definitions of governance, they have in common that they refer to governing styles in which the boundaries between and within public and private responsibilities have become blurred (e.g. Rhodes, 2007; Stoker, 1998).

When it comes to dealing with public issues in the environmental policy domain, many scholars claim that a governance approach is needed in which responsibilities are shared between public and private actors. That is because environmental issues are troubled by uncertainties, complexities, and ambiguities (e.g. Driessen et al., 2012; Lemos & Agrawal, 2006). Likewise, scholars of climate adaptation governance argue that the 'wickedness' of climate adaptation induces new governance arrangements with more involvement of private actors such as citizens and businesses (e.g. Lorenzoni et al., 2007; Mees, 2017; Termeer et al., 2013, 2017). The divisions of responsibilities between public and private actors in climate adaptation have been conceptually explored in recent years (Mees, 2017; Mees et al., 2012).

Regardless of this normative strive towards *more* governance in climate adaptation, it is also argued that in specific instances precisely *more* government is needed. Governments as public actors have an important role to play in supporting and enabling climate adaptation at multiple levels (Urwin & Jordan, 2008). An important consideration for assuming responsibility with governments is *effectiveness*. In situations in which market failure leads to a complete lack of adaptation, maladaptation, or insufficient adaptation, governments may need to take up responsibilities to raise the effectiveness of adaptation action (Mees, 2017). In such cases governments can require citizens to take up insurance to cover the damages of extreme events; or they can become insurance providers (Aakre & Rübbelke, 2010; Mendelsohn, 2006; Osberghaus et al., 2010). Governments can also generate and distribute knowledge on climate impacts as public goods (Aakre & Rübbelke, 2010; Osberghaus et al., 2010; Stern, 2007), for instance in situations in which private actors do not have access to sufficient information on climate risks, impacts, and solutions. Governments can also ensure *equity*, by correcting for the distributional consequences of climate impacts and of adaptation action (Bulkeley et al., 2013; Marino & Ribot, 2012). Climate change leads to different impacts on different groups and localities (e.g. Hess, 2008), and governments can help out those groups and localities most affected by those climate impacts (Osberghaus et al., 2010; Stern, 2007). Governments are also regarded to be key actors in matters of national security, which includes water safety issues such as severe flooding (Mees, 2017). In most Western European countries governments are responsible for the construction and maintenance of flood defense and emergency planning (Aakre & Rübbelke, 2010; Heltberg et al., 2009; Osberghaus et al., 2010), in order to ensure an effective, legitimate, and equitable adaptation to increased flood risks.

In contrast, an important consideration for assuming responsibilities with *private* actors is *efficiency*: private action is regarded to be more efficient and innovative (Mendelsohn, 2006; Stern, 2007). For instance, insurance companies can stimulate the uptake of adaptive building measures to reduce the impacts of floods to private buildings through differentiation of insurance fees, or discourage building in flood plains (Mees, 2017). Another consideration that is often cited is that it raises the support for and *legitimacy* in terms of input, throughput, or output of a policy (Adger et al., 2009; Mees, 2017; Paavola, 2008). Public policy is viewed as more legitimate when the decision-making process is participatory and deliberative and involves both public and private actors (e.g. Dryzek, 2000; Smith, 2003). It is also claimed that such interactive policy making promotes joint fact-finding and social learning processes, thus raising the adaptive capacity of society to cope with climate change (Driessen et al. 2001; Gupta et al., 2010; Pahl-Wostl, 2009). Besides the list of possible positive effects of interactive policy making, there are also many authors who point out the possible pitfalls of (too much) participation (e.g. Newig et al., 2018; van der Heijden & Ten Heuvelhof, 2012). For example, the possible exponential costs in terms of time and money, and the overrepresentation of small groups.

In the section "A Framework for 'Thick' Analysis" the aforementioned considerations for assuming public and/or private responsibilities, i.e. effectiveness, equity, efficiency, and legitimacy, will be used as evaluation criteria to conduct an analysis of the responsibilization of citizens in climate change adaptation in the Netherlands.

Citizen Responsibilization in Climate Adaptation Governance

In the debate about public and private responsibilities in the governance of climate adaptation, the role of citizens and communities gains increasing attention (Hegger et al., 2017). Governments involve citizens through participation and collaboration in decision-making processes of issues that directly affect them, and through the coproduction of public services. These developments are closely linked to the rise of bottom-up initiatives in which consumers create innovative solutions to become more self-supportive in terms of energy, water, or food production (De Vries et al.,

2016). Citizens are seen as empowered actors with resources who can contribute to the resilience of their communities, as is for instance propagated by the *Big Society* program in the United Kingdom. In the Netherlands, a similar political agenda of the *Energetic Society* was introduced in 2011. The premise is that the energy and creativity of citizens is both desirable and much needed to solve complex societal issues in addition to governmental action (Hajer, 2011). Citizens are encouraged to initiate all kinds of community initiatives such as for instance in the area of community care, urban green maintenance, and renewable energy collectives to name a few (Hajer, 2011; Tonkens, 2014). Citizens are also crucial actors for realizing adaptation measures in and around the house. Citizens' initiatives or consent is often necessary (Mees et al., 2012; Tompkins & Eakin, 2012) while they can also play a role in tailoring adaptation measures in terms of technical (im) possibilities, specificities of climate risks and residents' individual needs (Wamsler & Brink, 2014).

This results in an increased responsibilization of citizens (Klein et al., 2017; O'Hare et al., 2016; Roth & Prior, 2014; Wamsler, 2016) in which the government encourages the governed to become responsible for issues previously held to be the responsibility of government authorities (Barry et al., 1996: 29). The government enables, persuades, entices, or nudges citizens to 'take responsibility' for their lives and their communities (Peeters, 2013: 584). The responsibilization of citizens in climate change adaptation is discussed by both scientists and policymakers. Hegger et al. (2017) distinguish between three types of roles of residents in climate change adaptation: (1) as citizens vis-à-vis the government, (2) as consumers vis-à-vis the market, and (3) as civil society members.

Such a distinction is useful, as it provides directions for how these citizens' roles can be promoted and stimulated through government interventions: an increased responsibilization of citizens also requires a shift of roles from the side of the government: from a steering and regulating government, to a facilitating and enabling government (e.g. Gilbert, 2005) that supports, rather than directs citizens. To stimulate the roles of residents as citizens, governments will need to engage citizens more actively and on an equal basis, so as to empower them to take on those types of responsibilities (Hegger et al., 2017). As such, governments employ a "subtle way of 'stepping into' society and managing citizen behavior" (Peeters, 2013: 585). Governments can, for instance, provide financial incentives for individual adaptation; and develop formal or informal agreements with individuals who engage in improving city–citizen collaboration (Wamsler, 2016). To enhance the roles of citizens as consumers vis-à-vis the market, governments can also create or regulate markets for adaptation products through the use of taxes and subsidies for entrepreneurs or for citizens (Hegger et al., 2017). To promote the roles of citizens as civil society members, governments merely need to have a facilitating role. Often, these citizens' activities will be bottom-up forms of self-governance which by definition are not orchestrated by governmental actors, and therefore an adaptive and receptive stance towards such initiatives is necessary (Edelenbos et al., 2017). Facilitation can be done in various forms, such as for instance by establishing knowledge-sharing dialogues, schooling and other forms of citizen empowerment, and allowing for experimentation by providing legal exemptions or financial support (Hegger et al., 2017; Wamsler, 2016).

A Framework for a 'Thick' Analysis

Adger et al. (2003) have coined the term 'thick analysis' to provide an alternative to taking a sectoral view to environmental issues that leads to 'thin' explanations. Hence, a thick analysis includes multiple indicators in order to provide insight in the different values that affect environmental decisions. As illustrated in the previous section and supported by Adger et al. (2003) (but also Haus et al. 2004 and Kemp et al. 2005), such indicators are efficiency, effectiveness, equity and legitimacy. Furthermore, Adger et al. (2003) point out that while applying a thick analysis the physical, social, and institutional context matter and should be taken into consideration. We will briefly conceptualize these four indicators below before we apply them to the Dutch case study (also see Table 10.1). At the beginning of the case study, we will address the contexts influencing local climate adaptation in the Netherlands.

Effectiveness refers to the capacity of a decision to achieve its expressed objectives (Adger et al., 2003). Effectiveness can be measured in terms of (financial) costs and benefits, or by purely looking at the extent to which there is a match between predefined goals and actual outcomes. However, this might not be a sufficient way to look at goals, as goals can be unrealistic, contested between stakeholders, or set without any underlying problem. Therefore, it is important to measure the level of support for the goals and the outcomes. This indicates that there is a strong link between effectiveness and legitimacy.

Efficiency relates to the use and allocation of resources towards environmental decisions. It highlights the relation between benefits and expenses and the emphasis generally lies on welfare maximization. In other words, efficiency takes stock of whether the specific goals are achieved in a cost-effective manner. Efficiency has become a dominant criterion in public policy-making more generally (Peters & Pierre, 1998), in environmental policy (Lemos & Agrawal, 2006) and in urban adaptation to climate change as a new field of environmental policy (Mees, 2014). Nevertheless, the emphasis on economic efficiency as the main selection criterion in decisions regarding environmental policy has become increasingly criticized

Table 10.1 *Operationalization of concepts*

Concept	Operationalization	Citizen responsibilization in local climate adaptation, and its effect on sustainable development
Effectiveness	The capacity of a decision to reach its expressed objectives	Adapting public and private space to climate risks – and not solely public space
Efficiency	The use of resources in relation to welfare maximization	Sharing costs/investments in adapting the city to climate change
Equity	The distributive consequences of a decision (distributive justice)	Not all citizens will be able to invest in adaptation measures and this can translate into unequal distribution of adaptation measures. But it can also translate into inequity, as less wealthy people who cannot invest in adaptation will become more vulnerable to flooding.
Legitimacy	The extent to which decisions are acceptable to participants (procedural justice)	Citizens vary in their acceptance of taking up more responsibilities; governments also differ in their acceptance to shift responsibilities towards citizens.

Source: Adapted from Adger et al. (2003).

because it neglects distributive justice issues related to, for instance, the allocation of costs and benefits (Bromley & Paavola, 2002; Lemos & Agrawal, 2006).

The concept of *Equity* focuses on distributive justice or the distributive consequences of environmental decisions (Adger et al., 2003). This refers to a fair distribution of costs, resources, and benefits, but also spatial impacts and/ or political change that occurs due to environmental impacts. Fair, however, does not always mean equal. As Adger et al. (2003) point out, 'sometimes equity may require distribution according to contribution, whereas at other times need or equality may be the most appropriate basis for equitable decisions in terms of their outcomes' (also argued by Bromley & Paavola, 2002; Radin, 1996).

Legitimacy refers to the acceptance of the authority, and with that the support for their decisions, by the people (Bernstein, 2005; Biermann & Gupta, 2011; Mees et al., 2014;). This is also referred to as procedural justice. The extent of acceptance is often measured in terms of values, norms, and rules (Beetham, 1991). Legitimacy can be gained through engaging and participating with stakeholders throughout the environmental decision-making process. Inclusiveness (input) and deliberation (throughput) during the decision-making process can increase the support for the decisions and the actual outcome (output).

Adapting Dutch Cities with the Help of Citizens

A retrospective thick analysis is here applied on the Netherlands. For this, we have gathered illustrative examples from self-organized workshops with Dutch municipalities and regional water authorities; in-depth interviews with local policy makers from various Dutch municipalities; and involvement in the *City Deal* program on climate adaptation (in Dutch: *Agenda Stad*). The City Deal is a collaboration among national government, municipalities, and stakeholders to stimulate growth, innovation and livability of Dutch cities (Agenda Stad, 2017). There is collaboration in various relevant themes, among which climate adaptation. Municipalities interested in this topic can participate in an exchange of information and in this way learn from each other. Some of the authors of this chapter have participated in the *City Deal* 'climate adaptation and social initiatives' and provided reflections upon the sessions organized on this topic. The overall issues discussed in these sessions related to how municipalities can facilitate citizen initiatives and how to stimulate initiatives in less active neighbourhoods. Together these illustrative examples provide a prefiguration of how citizen responsibilization in local climate adaptation could work.

While the Netherlands can expect a variety of climate change risks such as floods and heat stress, the main focus has been on floods (Hegger et al., 2017; Runhaar et al., 2012). Almost two thirds of the country is susceptible to flooding. Therefore, flood defense, with a large system of dikes, dunes, barriers, and sluices, has historically been very important, and still continues to be to date (Gralepois et al., 2016; Kaufmann et al., 2016). Tasks related to the defense against flood risks from rivers and sea is implemented by both 'Rijkswaterstaat' (the Dutch office of Public Works) and 26 regional water authorities, both functionally specialized agencies that operate in relative isolation from political whims (Kaufmann et al., 2016). Responsibilities for the governance of pluvial flooding lie with the municipalities and its citizens. Municipalities are responsible for the efficient collection and processing of rainwater run-off on public grounds, while citizens are responsible for collecting rainwater on their own properties. Yet, flood risks from heavy rainfall have predominantly been tackled by using and increasing the capacity of the sewage system. As a result of climate change, flood incidents from heavy rainfall are occurring more frequently and have resulted in considerable material damage and regular inundations of infrastructure (NRC, 2016). Municipalities increasingly call upon their citizens to contribute to mitigating risks from pluvial flooding. Citizens are, for instance, required to store a certain amount of rainwater on their properties (Volkskrant, 2017).

Effectiveness

The general objective of Dutch cities is to adapt their cities to heavy rainfall events in order to allow none or little[1] urban flooding. Local governments currently have the responsibility to collect and process excessive rainwater by adapting public space. But solely adapting public space might not be sufficient to meet the expressed objective. In order to reduce vulnerability to urban flooding both public and private space needs to be adapted. For this, citizen responsibilization in local climate adaptation sounds as a promising solution. Private property owners should also take adaptation measures or at least, be made aware that they should not install measures that lead to maladaptation, for example, by paving gardens which reduces infiltration of storm water. Clear communication and knowledge sharing regarding what measures will facilitate local climate adaptation is crucial. Overall, shifting responsibilities towards citizens can assist in raising awareness of and accelerate the investments in local climate adaptation throughout a city. Hence, it could be stated that in terms of effectiveness, citizen responsibilization seems logical. Examples in which local governments invest in communication on the matter are the programs 'Amsterdam Rainproof' and 'Utrecht Waterproof'. In these programs, local governments aim to establish a network for citizens, local businesses, and other stakeholders to share knowledge and resources (Uittenbroek et al., 2014). These programs show results in terms of increasing awareness for the topic. But it is too early to state whether these programs are effectively contributing to the reduction of the vulnerability of an area to urban flooding.

Efficiency

As stated before, local governments are generally taking the lead in local climate adaptation (Bulkeley et al., 2013; Mees et al., 2012). Some are investing in physical measures such as water squares, green roofs, and expanding green infrastructure, as is the case in Rotterdam, Tiel, and Den Bosch. Other cities such as Amsterdam and Utrecht are setting up networks in which social learning and involving of citizens and local organizations are key. In general, cities acknowledge the urgency to address climate change by adapting the urban design. Yet, finding resources to invest solely in adaptation measures is a difficult task. By mainstreaming climate adaptation in existing urban policies, they intend to share resources and deal with

[1] Allowing streets to have water on them for a couple of hours is allowed.

multiple policy goals (Uittenbroek et al., 2013). For example, the water squares in Tiel and Rotterdam function not only as water storage facilities but also as public places for encounters and play. This illustrates that local governments are looking for efficient ways to allocate resources to local climate adaptation. Reallocating responsibilities to citizens is another way to increase efficiency. Citizens can assist by removing pavement in private gardens or by collecting and storing excessive rain water on private property (e.g. in rain barrels or green roofs) before discharging it to the public sewage system. Local governments can share their responsibilities for the collection and discharge of water, and in this way, possibly reduce investments in measures in public space. Besides actively adapting their own property, several Dutch municipalities also wish to have citizens involved and participating in the design, implementation, and managing of adaptation measures in public space. Several municipalities are exploring whether citizens want to take responsibility for maintenance tasks of, for example, public greens. Yet, some municipalities have experienced that citizens do not yet see this as their task. Citizens generally like to have influence on the design, but not necessarily on maintenance tasks which require a frequent and consistent time contribution over a longer time period. There are few examples in which citizens do want to take up these responsibilities for maintenance, but generally these concern very small lots. In terms of efficiency, citizen responsibilization in local climate adaptation sounds useful, as costs and benefits can be shared between public and private actors, and within policy domains as adaptation solutions such as water squares and additional green infrastructure generally also add to the esthetic quality of the built environment, health, biodiversity, etc. (Gill et al., 2007).

Equity

Citizen responsibilization can thus be an efficient way to address local climate adaptation. Yet, there might be equity issues. If all citizens took their newly assigned responsibility for adapting their properties to climate change, this would lead to fairness and equity. However, not all citizens might have the resources (time, finances, and knowledge) to invest in adaptation measures. For example, Mees et al. (2016) found examples in which it was mostly the highly educated people who took part. These people have the knowledge, access to government networks, and most likely resources to invest in adaptation measures. Citizens who know how to organize themselves and know how to present ideas to the local government are most likely to get attention and assistance of the municipality. However, there will also be groups of citizens who do not have these characteristics, which makes it more difficult for them to take up responsibilities for local climate adaptation. During meetings with various Dutch municipalities, several municipalities said to be reluctant to shifting responsibilities to citizens because they are afraid of unequal distribution of climate adaptation measures. Those who are capable of bearing the costs will invest in sufficient measures, yet those who are not will fall behind and become more vulnerable to urban flooding. This will fuel inequity. The local policy makers aim to have an even allocation of adaptation resources throughout the city. However, they also recognize that currently only a select group of citizens requests for facilitation of adaptation initiatives or is capable of taking adaptation measures on their own property. Therefore, equity will most likely become problematic if the Dutch municipalities continue to shift responsibilities to citizens as inequity can come from citizen responsibilization. Yet, our examples show that many Dutch municipalities are also aware of and have experienced these differences between citizens. This makes them reluctant to simply shift these responsibilities without seeking ways to facilitate/empower all kinds of citizens.

Legitimacy

As indicated before effectiveness relates to legitimacy in which citizens accept the decisions made by public authorities. In this analysis, the question is whether citizens will accept the shift in responsibilities for local climate adaptation. While from the perspective of effectiveness this could be considered a legitimate shift, in practice, this shift is not easily legitimized. Citizens have different interests and consequently, vary in the amount of responsibility they want to have. In the Rooftop Park Rotterdam project, a group of citizens wanted to participate in the planning, implementation and maintenance of the park. But the municipality only gave them limited responsibilities, because the maintenance department of the municipality was afraid that citizens would not uphold their responsibilities over a longer period of time. As opposed to that, in Amsterdam the municipality was willing to hand over maintenance responsibilities of an adapted public square to citizens from the neighborhood, but citizens considered maintenance a government task. These two examples illustrate two relevant findings: (1) it takes two parties to legitimize the shift in responsibilities – both citizens and governments need to accept this decision of citizen responsibilization in local climate adaptation; and (2) both parties rely on historical institutions: the government as main responsible actor in addressing public issues. As a result, citizens expect local government to take their responsibilities in reducing vulnerability. They also expect that governments take their responsibilities seriously and that they want to circumvent situations of which unsatisfied citizens arise. In order for citizen responsibilization in local climate adaptation to be successful, this shift in responsibilities needs to be considered legitimate by most stakeholders. While society in general can benefit from reduced vulnerability to urban flooding, not all citizens might see the direct need to address local climate adaptation. Hence, citizens might not accept this shift in responsibilities and accordingly, might not take up these responsibilities.

Conclusion

Worldwide, there is an increasing focus on local climate adaptation. Currently, cities are learning how complex this adaptation challenge is. The issue cannot be solved solely by local governments, but requires the involvement of citizens. We have shown, however, that Western European countries such as the Netherlands are struggling with how to involve and share responsibilities with citizens in this challenge of local climate adaptation. As illustrated in our research, increased citizen responsibilization could have benefits in terms of effectiveness and efficiency, as this might stimulate investments in adaptation measures in private space and the sharing the costs for local climate adaptation with multiple stakeholders. But at the same time, there are good reasons to assume that citizen responsibilization increases inequalities and might lead to legitimacy problems in some circumstances. To begin with, the actual handing over of responsibilities to citizens may be problematic. In our Dutch examples, we have seen cases in which authorities are reluctant to actually hand over responsibilities. In other cases, however, the challenge might be that citizens do not accept the fact that administrations give down responsibilities rather than dealing with them themselves. Once responsibilities have actually been delegated to citizens, other issues might arise. Not all citizens will be capable of or able to invest in adaptation measures in private space. They might not have the knowledge, resources, or networks. At the same time, other citizens might not consider local climate adaptation their responsibility or do not consider it in their interest to make personal investments in climate adaptation. In other words, they might not legitimize this shift in responsibilities from the public to the private domain. If citizens are not capable of implementing adaptation measures or are not willing to accept new responsibilities, the required level of local climate adaptation may not be reached. This is problematic as this can lead to maladaptation or (increased) vulnerability to climate risks. This can be read as an argument to tailor efforts at citizen responsibilization to different target groups. Loosely based on Dahl (1989), one can argue that to provide equal opportunities to participate, groups need to be treated differently.

The trend of citizen responsibilization will most likely continue and therefore it is relevant to consider ways to deal with the tensions between these four criteria. We consider three (intertwined) pathways for this. First, the focus should be on processes of *upscaling of local adaptation initiatives*. In spite of the fact that there might be citizens who are unwilling or unable to invest in climate adaptation, there are also citizens who are already adapting their own properties or who are participating in social initiatives that relate to climate adaptation. Understanding these good practices and deriving lessons from them is relevant and useful for gaining ground. These lessons can be crucial to stimulate both *horizontal* and *vertical* upscaling (van Doren et al., 2016). *Horizontal* upscaling refers to the possibility of using lessons as input for copying successes. *Vertical* upscaling refers to structural learning and changing existing institutions to support local climate adaptation by citizens. We hold that the learning that provides input to these upscaling processes should pertain to all of the four considerations discussed earlier: effectiveness, efficiency, equity, and legitimacy. This relates to our second pathway, *improving interactions between governments and citizens*. For each country, the relation between governments and its citizens might vary widely. In many Western European countries, this relationship is based on trust and solidarity; while in other countries, citizens expect governments to do as little as possible as is the case in the United States and the United Kingdom. Facing climate change is however a problem that requires 'all hands on deck'. Governments can play a guiding, stimulating, and/or facilitating role in preparing cities for climate change risks. As an example, Amsterdam does this by establishing a network in which knowledge, resources, and connections are shared (Bulkeley et al., 2013). This brings us to the final pathway in which governments need to *gain experiences with how to substantiate this 'facilitating' role that is often proclaimed in literature* (Hegger et al., 2017). In order to improve the relationship with citizens and to learn how to vertically upscale the citizen responsibilization, governments have to experiment with policy instruments. Communication through setting up such networks and providing subsidies for adaptation measures are possible ways to facilitate citizens in local climate adaptation (Mees et al., 2014).

Still, the question remains if mobilizing citizens in the end leads to sufficient capacity to solve the anticipated problems (see also van der Heijden & Heuvelhoff, 2012). Not many cities choose to regulate climate adaptation. However, it can be questioned if non-committal approaches in the end will lead to the mobilization of sufficient adaptive capacity. Possible explanations for the non-committal character of citizen responsibilization to date are that (1) local governments want citizen responsibilization but do not know or agree to what extent; or (2) they might not know how to enforce these responsibilities. There is also a fine line between 'dumping' responsibilities on citizens and empowering citizens to take up their responsibilities. Some governments are aware of this and therefore experiment with more soft steering instruments (Bulkeley & Castán Broto, 2013). Yet, if we want to keep our cities liveable, local climate adaptation cannot go without obligations (Mees, 2017; Runhaar et al., 2016).

References

Aakre, S., & Rübbelke, D. T. G. (2010). Objectives of public economic policy and the adaptation to climate change. *Journal of Environmental Planning and Management*, 53(6): 767–791.

Adger, W. N., Arnell, N. W., & Tompkins, E. L. (2005). Successful adaptation to climate change across scales. *Global Environmental Change*, 15(2): 77–86.

Adger, W. N., Brown, K., Fairbrass, J., et al. (2003). Governance for sustainability: Towards a 'thick' analysis of environmental decision making. *Environment and Planning A: Economy and Space*, 35(6): 1095–1110.

Adger, W. N., Dessai, S., Goulden, M., et al. (2009). Are there social limits to adaptation to climate change? *Climatic Change*, 93, 335–354.

Agenda Stad. (2017). Agenda Stad. www.agendastad.nl

Arts, B. (2014). Assessing forest governance from a 'Triple G'perspective: Government, governance, governmentality. *Forest Policy and Economics*, 49: 17–22.

Barry, A., Osbome, T., & Rose, N., eds. (1996). *Foucault and Political Reason: Liberalism, NeoLiberalism and Rationalities of Government*. London: University College Press.

Beetham, D. (1991). Max Weber and the legitimacy of the modern state. *Analyse & Kritik*, 13(1): 34–45.

Bernstein, S. (2005). Legitimacy in global environmental governance. *Journal of International Law and International Relations*, 1: 139–166.

Biermann, F., & Gupta, A. (2011). Accountability and legitimacy in earth system governance: A research framework. *Ecological Economics*, 70: 1856–1864.

Bromley, D. W., & Paavola, J. (2002). Economics, ethics and environmental policy. In D. W. Bromley & J. Paavola (eds.), *Economics, Ethics, and Environmental Policy: Contested Choices*. Oxford: Blackwell Publishers, 261–276.

Bulkeley, H., Carmin, J., Castán Broto, V., & Edwards, G. A. S. (2013). Climate justice and global cities: Mapping the emerging discourses. *Global Environmental Change*, 23: 914–925.

Bulkeley, H., & Castán Broto, V. (2013). Government by experiment? Global cities and the governing of climate change. *Transactions of the Institute of British Geographers*, 38 (3): 361–375.

Capano, G., Howlett, M., & Ramesh, M. (2015). Bringing governments back in: Governance and governing in comparative policy analysis. *Journal of Comparative Policy Analysis: Research and Practice*, 17(4): 311–321.

Colvin, C. J., Robins, S., & Leavens, J. (2010). Grounding 'responsibilisation talk': Masculinities, citizenship and HIV in Cape Town, South Africa. *Journal of Development Studies*, 46(7): 1179–1195.

Dahl, R. A. (1989). *Democracy and Its Critics*. New Haven, CT: Yale University Press.

De Vries, G., Boon, W., & Peine, A. (2016). User-led innovation in civic energy communities. *Environmental Innovation and Societal Transitions*, 19: 51–65.

Driessen, P. P. J., Dieperink, C., van Laerhoven, F., Runhaar, H. A. C., & Vermeulen, W. J. V. (2012). Towards a conceptual framework for the study of shifts in environmental governance: Experiences from the Netherlands. *Environmental Policy and Governance*, 22(3): 143–160.

Driessen, P. P. J., Glasbergen, P., & Verdaas, C. (2001). Interactive policy-making: A model of management for public works. *European Journal of Operational Research*, 128: 322–337.

Driessen, P. P. J., & van Rijswick, H. F. M. W. (2011). Normative aspects of climate adaptation policies. *Climate Law*, 2(4): 559–581.

Dryzek, J. S. (2000). *Deliberative Democracy and Beyond: Liberals, Critics, Contestations*. Oxford: Oxford University Press.

Dubbink, W. (2003). *Assisting the Invisible Hand: Contested Relations between Market, State and Civil Society*. Dordrecht: Kluwer Academic.

Edelenbos, J., van Buuren, A., Roth, D., & Winnubst, M. (2017). Stakeholder initiatives in flood risk management: Exploring the role and impact of bottom-up initiatives in three 'Room for the River' projects in the Netherlands. *Journal of Environmental Planning and Management*, 60(1): 47–66.

Gilbert, N. (2005). The "Enabling State?" From Public to Private Responsibility for Social Protection.

Gill, S. E., Handley, J. F., Ennos, A. R., & Pauleit, S. (2007). Adapting cities for climate change: The role of the green infrastructure. *Built Environment*, 33(1): 115–133.

Gralepois, M., Larrue, C., Wiering, M., et al. (2016). Is flood defense changing in nature? Shifts in the flood defense strategy in six European countries. *Ecology and Society*, 21 (4): 37.

Gupta, J., Termeer, K., Klostermann, J., et al. (2010). The Adaptive Capacity Wheel: A method to assess the inherent characteristics of Institutions to enable the adaptive capacity of society. *Environmental Science & Policy*, 13: 459–471.

Hajer, M. (2011). The energetic society: In search of a governance philosophy for a clean economy. Report. The Hague: PBL Netherlands Environmental Assessment Agency.

Haus, M., Heinelt, H., & Stewart, M., eds. (2004). *Urban Governance and Democracy: Leadership and Community Involvement*. London: Routledge.

Hegger, D. L. T., Driessen, P. P. J., Dieperink, C., Wiering, M., Raadgever, G. T. T., & van Rijswick, H. F. M. W. (2014). Assessing stability and dynamics in flood risk governance: An empirically illustrated research approach. *Water Resources Management*, 28 (12): 4127–4142.

Hegger, D. L. T., Mees, H. L. P., Driessen, P. P. J., & Runhaar, H. A. C. (2017). The roles of residents in climate adaptation: A systematic review in the case of the Netherlands. *Environmental Policy and Governance*, 27(4): 336–350.

Heltberg, R., Siegel, P. B., & Jorgensen, S. L. (2009). Addressing human vulnerability to climate change: Toward a 'no-regrets' approach. *Global Environmental Change*, 19: 89–99.

Hess, J. J., Malilay, J. N., & Parkinson, A. J. (2008). Climate change: The importance of place. *American Journal of Preventive Medicine*, 35(5): 468–478.

Ilcan, S., & Basok, T. (2004). Community government: Voluntary agencies, social justice, and the responsibilization of citizens. *Citizenship Studies*, 8(2): 129–144.

Jordan, A., & Lenschow, A. (2010). Policy paper environmental policy integration: A state of the art review. *Environmental Policy and Governance*, 20(3): 147–158.

Jordan, A., Wurzel, R. K., & Zito, A. (2005). The rise of 'new' policy instruments in comparative perspective: Has governance eclipsed government? *Political Studies*, 53 (3): 477–496.

Kaufmann, M., Doorn-Hoekveld, W. V., Gilissen, H. K., & van Rijswick, H. F. M. W. (2016). Analysing and evaluating flood risk governance in the Netherlands: Drowning in safety. Report. Utrecht: STARFLOOD Consortium.

Kemp, R., Parto, S., & Gibson, R. B. (2005). Governance for sustainable development: Moving from theory to practice. *International Journal of Sustainable Development*, 8 (1–2): 12–30.

Kistner, U. (2009). Adversities in adherence: Paralogisms of 'Biological Citizenship' in South Africa. Paper presented at the European Conference of African Studies, 4–7 June, Leipzig.

Klein, J., Juhola, S., & Landauer, M. (2017). Local authorities and the engagement of private actors in climate change adaptation. *Environment and Planning C: Politics and Space*, 35(6): 1055–1074.

Lemos, M. C., & Agrawal, A. (2006). Environmental governance. *Annual Review of Environment and Resources*, 31: 297–325.

Lorenzoni, I., Jones, M., & Turnpenny, J. R. (2007). Climate change, human genetics, and post-normality in the UK. *Future*, 39: 65–82.

Marino, E., & Ribot, J. (2012). Special issue introduction: Adding insult to injury: climate change and the inequities of climate intervention. *Global Environmental Change*, 22 (2): 323–328.

Mees, H. (2017). Local governments in the driving seat? A comparative analysis of public and private responsibilities for adaptation to climate change in European and North-American cities. *Journal of Environmental Policy and Planning*, 19(4): 374–390.

Mees, H. L. P., Driessen, P. P. J., & Runhaar, H. A. C. (2012). Exploring the scope of public and private responsibilities for climate adaptation. *Journal of Environmental Policy and Planning*, 14(3): 305–330.

Mees, H. L., Driessen, P. P., & Runhaar, H. A. (2014). Legitimate adaptive flood risk governance beyond the dikes: The cases of Hamburg, Helsinki and Rotterdam. *Regional Environmental Change*, 14(2): 671–682.

Mees, H., & Uittenbroek, C. J. (2016). Dutch Country report Transadapt.

Mendelsohn, R. (2006). The role of markets and governments in helping society adapt to a changing climate. *Climatic Change*, 78: 203–215.

Newig, J., Challies, E., Jager, N. W., Kochskaemper, E., & Adzersen, A. (2018). The environmental performance of participatory and collaborative governance: A framework of causal mechanisms. *Policy Studies Journal*, 46(2): 269–297.

NRC. (2016). Buien met veel regen komen nu vaker voor. Klimaatverandering is niet per definitie de oorzaak, 3 juni.

O'Hare, P., White, I., & Connelly, A. (2016). Insurance as maladaptation: Resilience and the 'business as usual' paradox. *Environment and Planning C: Government and Policy*, 34(6): 1175–1193.

Olsson, L., Opondo, M., Tschakert, P., et al. (2015). Livelihoods and poverty. In *Climate Change 2014: Impacts, Adaptation, and Vulnerability. Part A: Global and Sectoral Aspects*. Contribution of Working Group II to the Fifth Assessment Report of the Intergovernmental Panel on Climate Change [Field, C. B., V. R. Barros, D. J. Dokken, K. J. Mach, M. D. Mastrandrea, T. E. Bilir, M. Chatterjee, K. L. Ebi, Y. O. Estrada, R. C. Genova, B. Girma, E. S. Kissel, A. N. Levy, S. MacCracken, P. R. Mastrandrea, and L. L. White (eds.)]. Cambridge University Press, Cambridge, United Kingdom and New York, NY, USA, 793–832.

Osberghaus, D., Dannenberg, A., & Mennel, T. (2010). The role of the government in adaptation to climate change. *Environment and Planning C: Government and Policy*, 28: 834–850.

Paavola, J. (2008). Science and social justice in the governance of adaptation to climate change. *Environmental Politics*, 17(4): 644–659.

Pahl-Wostl, C. (2009). A conceptual framework for analysing adaptive capacity and multi-level learning processes in resource governance regimes. *Global Environmental Change*, 19: 354–365.

Peeters, R. (2013). Responsibilisation on government's terms: New welfare and the governance of responsibility and solidarity. *Social Policy and Society*, 12(4): 583–595.

Peters, B. G., & Pierre, J. (1998). Governance without government? Rethinking public administration. *Journal of Public Administration Research and Theory*, 8(2): 223–243.

Radin, M. J. (1996). *Contested Commodities*. Cambridge, MA: Harvard University Press.

Rhodes, R. A. (2007). Understanding governance: Ten years on. *Organization Studies*, 28 (8): 1243–1264.

Rockström, J., Steffen, W., Noone, K., et al. (2009). A safe operating space for humanity. *Nature*, 461(7263): 472–475.

Roth, F., & Prior, T. (2014). The boundaries of building societal resilience: Responsibilisation and Swiss Civil defense in the Cold War. *Behemoth: A Journal on Civilisation*, 7(2): 103–123.

Runhaar, H., Mees, H., Wardekker, A., van der Sluijs, J., & Driessen, P. P. J. (2012). Adaptation to climate change-related risks in Dutch urban areas: Stimuli and barriers. *Regional Environmental Change*, 12(4): 777–790.

Runhaar, H. A. C., Uittenbroek, C. J., van Rijswick, H. F. M. W., Mees, H. L. P., Driessen, P. P. J., & Gilissen, H. K. (2016). Prepared for climate change? A method for the ex-ante assessment of formal responsibilities for climate adaptation in specific sectors. *Regional Environmental Change*, 16(5): 1389–1400. http://dx.doi.org/10.1007/ s10113-015-0866-2

Smith, G. (2003). Deliberative democracy and green political theory. In G. Smith (ed.), *Deliberative Democracy and the Environment*. New York: Routledge, 53–76.

Stern, N. (2007). *The Economics of Climate Change: The Stern Review*. Cambridge: Cambridge University Press.

Stoker, G. (1998). Governance as theory: Five propositions. *International Social Science Journal*, 50: 17–28.

Termeer, C., Dewulf, A., & Breeman, G. (2013). Governance of wicked climate adaptation problems. In J. Knieling & W. Leal Filho (eds.), *Climate Change Governance*. Berlin Heidelberg: Springer, 27–39.

Termeer, C., van Buuren, A., Dewulf, A., et al. (2017). Governance arrangements for adaptation to climate change. In *Oxford Research Encyclopedia of Climate Science*. Oxford: Oxford University Press.

Tompkins, E. L., & Eakin, H. (2012). Managing private and public adaptation to climate change. *Global Environmental Change*, 22(1): 3–11.

Tonkens, E. (2014). Vijf misvattingen over de participatiesamenleving. Afscheidsrede Universiteit van Amsterdam.

Uittenbroek, C. J., Janssen-Jansen, L. B., & Runhaar, H. A. C. (2013). Mainstreaming climate adaptation into urban planning: Overcoming barriers, seizing opportunities and evaluating the results in two Dutch case studies. *Regional Environmental Change*, 13(2): 399–411.

Uittenbroek, C. J., Janssen-Jansen, L. B., Spit, T. J., Salet, W. G., & Runhaar, H. A. C. (2014). Political commitment in organising municipal responses to climate adaptation: The dedicated approach versus the mainstreaming approach. *Environmental Politics*, 23(6): 1043–1063.

Urwin, K., & Jordan, A. (2008). Does public policy support or undermine climate change adaptation? Exploring policy interplay across different scales of governance. *Global Environmental Change*, 18: 180–191.

van der Heijden, J., & ten Heuvelhof, E. (2012). The mechanics of virtue: Lessons on public participation from implementing the Water Framework Directive in the Netherlands. *Environmental Policy and Governance*, 22(3): 177–188.

van Doren, D., Driessen, P. P., Runhaar, H., & Giezen, M. (2018). Scaling-up low-carbon urban initiatives: Towards a better understanding. *Urban Studies*, 55(1): 175–194.

van Kersbergen, K. & van Waarden, B. F. (2001). *Shifts in Governance: Problems of Legitimacy and Accountability*. NWO Netherlands Organization for Scientific Research.

Volkskrant (2017). Straten vaker blank door extreme regenval en waarom u moet meebetalen aan de oplossing. 17 juli.

Wamsler, C. (2016). From risk governance to city-citizen collaboration: Capitalizing on individual adaptation to climate change. *Environmental Policy and Governance*, 26 (3): 184–204.

Wamsler, C., & Brink, E. (2014). Interfacing citizens' and institutions' practice and responsibilities for climate change adaptation. *Urban Climate*, 7: 64–91.

11

Agency and Climate Governance in African Cities

Lessons from Urban Agriculture

CHRISTOPHER GORE

Introduction

Nine cities in sub-Saharan Africa (Africa) are active members of the C40 Cities climate change network – 'a network of megacities committed to addressing climate change'. When commenting on the city of Dar es Salaam's participation in the C40, the city's Lord Mayor stated: 'Through C40's inter-city connectivity, I would like to ensure that we have the requisite capacity – both financial and human – to understand climate change issues, and implement adaptation and mitigation activities focused on making Dar es Salaam a sustainable, resilient and livable city' (Mwita, 2016). The participation of African[1] city mayors or managers in global networks or forums to address climate change is now commonplace. The Former Executive Director of the city of Kampala, Jennifer Musisi, for example, spoke at the international Climate Action Summit 2016 in Washington, DC, and articulated the steps the city of Kampala was taking to mainstream climate change in city actions (Okanya, 2016). Owing to the climate vulnerability and risks that these two cities face (see Gore, 2015), this engagement is laudable and important. But participation at these forums often masks the complex character of authority and agency in African cities; that is, the forums do not provide insights into how climate mainstreaming may come about or has come to be; who exercises influence over policy and decision-making in cities; and, most critically, whether city residents are able to shape or alter the policy and governance context in which they are situated (see Chapter 1 by van der Heijden, Bulkeley & Certomà in this book).

In the case of both the Lord Mayor of Dar es Salaam and the Executive Director of Kampala, their participation in global climate 'orchestration platforms' (van der

[1] The chapter uses the term 'Africa' to generalize about countries south of the Sahara. The author recognizes the problems resulting from this generalization, therefore, uses the term when summarizing general experiences of a high number of countries in the sub-continent, while aiming to be specific whenever possible.

Ven, Bernstein & Hoffmann, 2017) provides a false sense of how the city leaders' authority to act on climate change is exercised locally and their legitimacy domestically. In the case of Dar es Salaam, the Lord Mayor represents the Dar es Salaam City Council (DCC), which functions as a weak coordinating body for the Region of Dar es Salaam, or the de facto metropolitan area of Dar es Salaam. The DCC has no authority over the five, independent municipalities that make up the city of Dar es Salaam, and no independent financial capacity or revenue streams to provide incentives for the other municipalities to take coordinated action (Mhagama, 2016).

In the case of Kampala, the Executive Director of the city is not elected. She is the Director of the Kampala Capital City Authority (KCCA), an institution created by the national government when it took over the capital city. The Director of KCCA derives authority from the national government and not the electorate, and serves at the pleasure of the national government (see Gore & Muwanga, 2014). While local elections in Kampala continue, since the national government took over the city, the elected mayor of Kampala and the Executive Director have been in near continual tension. Given that 'addressing climate change in the city provokes fundamental political tensions over how and for whom environmental protection ... should be pursued' (Bulkeley, 2010: 15 in Gore, 2015: 207); that African cities are typically understood to have weak revenue sources and weak capacity; and that cities are often dependent on national and international financial support and authority, it is critical to understand how and if African cities can and do exercise agency that might shape or alter climate impacts and outcomes domestically.

This chapter examines a central theme and question posed by this book: in relation to urban environmental governance, how is agency exercised in African cities and what novel forms of agency have been observed? It further examines how the multiple layers of authority, control, institutions and practices that African cities experience – what Earth System Governance scholars refer to as the architecture governing ecological transformations (Habtezion et al., 2015) – shape potential climate actions and opportunities for action. Hence, the chapter focuses on two dimensions of the analytical pillars of Earth System Governance – agency and architecture. African cities are understood to have weak agency generally (Myers, 2011; Parnell & Simon, 2014) and weak capacity to act on climate change (Habtezion et al., 2015). Further, this weak agency is compromised by the complex architecture of international and national institutions that have a prominent influence over African cities due to their weak capacity and very limited financial resources (Stren, 2014). Yet, to assume African states and cities lack agency to navigate complex international and domestic political contexts is to dismiss a growing body of research that shows learning, engagement, and leadership in environmental governance and climate change internationally and domestically (Carmin, Anguelovski & Roberts, 2012; Gore, 2018; Roger & Belliethathan, 2016).

In examining the role and potential for 'bottom-up agency' in African city climate response, this chapter draws lessons from existing research on climate action in African cities and the response to another critical environmental and livelihood concern in African cities, food security. Lessons from city leadership in food security in East Africa, and city response to climate change more generally, shows how subnational policy leadership materializes and the lessons that can be drawn from this leadership for future climate adaptation. Drawing on secondary and primary research, the chapter emphasizes how ongoing, sustained dialogue between unelected members of local governments, citizens, and civil society provides the foundation for locally responsive environmental policy. This conclusion does not dismiss the need for financial and technical capacity and support, or the need for African central governments to better support cities (see Pieterse & Parnell, 2014). The goal is to recognize the fundamental political tensions that emerge in urban climate response (Bulkeley, 2010) and in African cities generally (Goodfellow & Titeca, 2012; Gore, 2015; Gore & Muwanga, 2014; Klaus & Paller, 2017; Paller, 2014; Resnick, 2012), and identify ways that these tensions have been and can be smoothed while responding to the critical climate impacts in African cities.

The chapter proceeds by first highlighting concerns with climate change and African cities, as well as their correlation with food security in cities. Here, the impacts of climate change are complemented by a reflection on how African national governments and cities have engaged in climate governance. The chapter then moves to examine the specific issue of climate and environmental policy leadership in East Africa. This section examines how policy leadership has emerged around the issue off food production in the region's largest cities. This section reveals two things: the correlation between climate and food security and how leadership in food security emerged, and how agency was exercised in the midst of the complex architecture cities find themselves. The conclusion draws lessons for African city climate response.

African States and Cities: Climate Impacts, Action, and Governance

The United Nations Development Program (UNDP) identifies 46 countries as part of sub-Saharan Africa. According to the World Bank, these countries have a population of just over 1 billion people, or roughly one eighth of the world population in 2017. Conversely, the total, combined carbon dioxide (CO_2)

emission contribution of the African continent as a whole, including major industrializing countries such as South Africa, Nigeria and all North African states, is less than individual contributions of the top four emitting countries in the world – China, the United States, India and Russia (Global Carbon Atlas, 2018; Peters et al., 2011). This gross discrepancy in contributions to global climate change has provoked challenging debates about international climate and energy justice (J. T. Roberts & Parks, 2007; Sovacool & Dworkin, 2014), but also questions about how global decarbonization and climate justice can be achieved at the same time as meeting the energy needs of African countries and cities. International organizations promote 'climate-compatible development' in African cities and countries; however, this goal is challenged by the 'complex and highly differentiated' energy trajectories and needs of African countries (Newell & Bulkeley, 2016: 5) and by the extraordinary demand for access to modern energy in African countries (Gore, 2017; Gore et al., 2018; MacLean et al., 2016; Newell & Mulvaney, 2013). Many scholars are engaging deeply with questions about how low-carbon, just energy transitions can and are being realized at the national and city and scales, including in Africa (Bulkeley et al., 2013; Bulkeley, Edwards & Fuller, 2014; Gore, 2017; Newell & Bulkeley, 2016; Newell & Mulvaney, 2013; Power et al., 2016). But in the midst of this research, the impacts of climate change on African countries and cities remain dramatic.

African countries and cities are expected to be severely affected by climate change (Niang et al., 2014; Revi et al., 2014). For a continent as large as Africa, it is impossible to generalize about climate change impacts. The impacts range from general water stress to increased average temperatures and heat extremes to greater intensity of rainfall events, to less predictable weather patterns as a whole. For food security and agricultural production, the impacts, not surprisingly, vary. Some 'worst-case' projections 'indicate losses of 27–32%' for staples such as maize, sorghum, millet, and groundnut, and temperatures already above the average for wheat (Serdeczny et al., 2017: 1591–1592). The relationship between food security and climate is not uniformly negative in the short term, however (de Waal, 2018). The Food and Agriculture Organization (FAO) notes that increased rainfall and CO_2 concentrations will stimulate plant growth (de Waal, 2018: 169). Past 2030, however, the negative impacts of climate change surpass any potential benefits, including for Africa (de Waal, 2018: 169). Further, the fact that drought may increase under climate change does not mean famine follows – famine is largely a result of the inadequate distribution of food supplies resulting from political choices or inadequate distribution systems. In general, climate change will cause greater instability in weather patterns, which can destabilize food availability and physical infrastructure, increase food prices, and contribute to social and political instability (see Bellemare, 2015; FAO, 2011).

At the city level in Africa, climate impacts are amplified in primate and secondary cities owing to poor infrastructure, particularly drainage, poor quality housing, unplanned settlements, and inadequate services. Flooding in low-lying cities is now routine throughout the continent. With increased flooding, poor services, and poor drainage, the potential for water contamination also increases, as well as less capacity to respond to risks and disasters (see Revi et al., 2014). The character of urban change in the subcontinent means that the urgency to respond to these challenges is intense. While the total percentage of urban residents in African cities is lower than in any other part of the world, urban populations in Africa are increasing faster than in any other region (Stren, 2014). Further, half of urban residents in the subcontinent are predicted to be living in poor quality, underserviced settlements past 2030 (Stren, 2014). With intensifying climate impacts, the historic and ongoing challenges of providing urban residents with services and infrastructure are amplified. The Intergovernmental Panel on Climate Change (IPCC), in its assessment of impacts for the African continent and for cities, refers to Africa's low adaptive capacity (Nieng et al., 2014; Revi et al., 2014): 'High levels of vulnerability and low adaptive capacity result from structural factors, particularly local governments with poor capacities and resources (Kithiia, 2011). Weak local government creates and exacerbates problems' (Revi et al., 2014: 1225). Put differently, 'The multi-dimensional complexities of urban form, urbanization and urban governance in Africa have left city authorities and governments unprepared for climate change' (Lwasa, 2010a: 20 in Gore, 2015: 210). The IPCC 2014 Assessment Report characterized the dilemma for African cities well: it is difficult to climate-proof infrastructure that does not exist, but hard infrastructural responses that may help address concerns are costly (Niang et al., 2014: 1225). Adding to this, hard infrastructure responses to climate change, particularly those imposed by external authorities, often underemphasize the conflict that these responses can and do generate (Bulkeley & Tuts, 2013). Hence, the impacts of climate change are recognized as dangerously problematic for human well-being in Africa. Yet research also generally predicts that large-scale, centralized, technical responses to mediate these risks are financially prohibitive and will likely produce political conflict in settings where tensions between city and national governments is commonplace (Goodfellow & Titeca, 2012; Gore & Muwanga, 2014; Klaus & Paller, 2017; Resnick, 2012), and where the capacity of local governments to respond is low (Revi et al., 2014).

These negative conditions generally give the impression that African state and city agency is weak. Yet, even if the resources were available to respond to climate impacts through international or national forums, there is an imperative to understand how the power dynamic between local, national, and international actors shapes and how these conditions can reproduce inequities and conflict in cities (Gore, 2015: 220–221). Or as Habtezion et al. (2015: 201) observe, 'Because adaptation takes place primarily at the local level, better understanding of the local-central power dynamics and its relationship to GEC [global environmental change] impacts needs to be better understood.' How then have African states and cities engaged in the architecture of climate governance globally and domestically?

African State and City Engagement in Climate Governance

Participation in formal, international environmental governance forums is a resource intensive, complex process, which requires a high degree of technical and strategic bargaining capacity and power to be able to leverage desired outcomes. The large number of independent African states and their diverse social, political, economic, and ecological conditions puts the subcontinent at a collective disadvantage in bargaining, as it is nearly impossible for governments to present a unified position in negotiations. In an analysis of African government engagement in the climate change regime for 'reducing emissions from deforestation and forest degradation' (REDD+), it was observed that African countries are deeply involved in the implementation of REDD+, but that the continent as a whole has weak agency in the design of the mechanism's architecture or structure: 'Africa's position is fragmented across negotiation coalitions which weakens the continent's influence on the REDD+ agenda' (Atela et al., 2017: 463–464). Following other Earth System Governance scholars, these authors define agency at the international level as the capacity 'to participate in negotiations and inform decisions within established norms' (Atela et al., 2017: 465). African states with large economies, such as South Africa, are often aligned with non-African states, such as Brazil or India, whose interests may run counter to those of a majority of poorer African states. These conditions have put African governments at a severe disadvantage independently and collectively in international climate regimes, exhibiting engagement but weak agency.

For climate change negotiations more generally, African states have also been characterized as having weak agency historically, although this is changing. Examining the African Group of Nations (AGN), Roger and Belliethathan (2016) note that 'During the early negotiations, the AGN suffered from a number of constraints, such as inadequate resources, limited access to high-quality information and poor negotiating skills, which made it difficult for African states to bargain effectively, both individually and as a group' (Roger & Belliethathan, 2016: 93). Yet more recently, states have used their 'symbolic power as a highly vulnerable group of states to exert pressure, staging walkouts and protests' and 'putting adaptation and CDM [clean development mechanism] capacity building at the center of talks' (Roger & Belliethathan, 2016: 92). Despite these outcomes, it remains that reliance on data from other countries, inexperienced delegates, and turnover in delegates (Roger & Belliethathan, 2016: 97–98) put African states at a severe disadvantage in the formal international climate governance system. This does not mean that African states are not well represented or engaged in climate change mitigation and adaptation planning. Under the UNFCCC process, 34 African least developed countries (LDCs) have adopted National Adaptation Programmes of Action (NAPAs), for example (Habtezion et al., 2015: 200). Adaptation, however, requires local governments to play lead roles. Yet, weak and tense local–central government relations in many African countries (see Goodfellow & Titeca, 2012; Gore, 2015, 2018; Gore & Muwanga, 2014; Paller, 2015), combined with multiple layers of formal and informal authority at different scales (see Habtezion et al., 2015: 199), mean that national state action in Africa cannot be understood without understanding how cities are also engaged.

African cities are participating actively in numerous experiments in subnational climate governance. As earlier noted, nine African cities are members of the C40 Cities climate network. Beyond this, one of the largest networks of African cities committed to climate action is the Global Covenant of Mayors for Climate and Energy (GCoM). The Covenant is a merger of the UN Compact of Mayors and the Covenant of Mayors; combined, in April 2018, the Covenant had 7,520 members (Global Covenant of Mayors, 2018). The vast majority of these are in Europe, but there are 75 members from the African continent. The Covenant also now has regional coordinating bodies, including one for Africa, titled the Covenant of Mayors in Sub-Saharan Africa (CoMSSA).

The value of subnational climate networks has been studied extensively, with researchers documenting the continued increase in membership and the benefits accrued from membership, particularly knowledge dissemination, networking, and demonstrated engagement with a global problem (Betsill & Bulkeley, 2003, 2004; Gore, 2010). At the same time, despite their continued increase in membership and popularity, it remains very difficult to assess the overall impact of these networks on reducing emissions, on motivating cities to reduce emissions, and whether network emphasis on measuring decarbonization is the best role for these 'orchestration platforms' (Kern & Bulkeley, 2009; Krause, 2012; Robinson & Gore, 2015; van der Ven et al., 2017). Many international networks have now increased their focus on producing actionable knowledge for cities to use in climate adaptation and resiliency planning. Indeed, four of the five reasons listed for joining the Covenant of Mayors for Sub-Saharan Africa (CoMSSA) are to gain knowledge, technical support, networking opportunities, and international visibility, with a fifth being to build a sustainable and resilient city (CoMSSA, 2018). Conversely, the first formal commitment of signatories to the CoMSSA Covenant is specifically tied to setting and achieving targets and undertaking initiatives (adaptation plans and staying in line with national commitments) that require direct engagement with the complex architecture of African states: The commitment states that cities will 'work towards setting ambitious targets for mitigation, designing adaptation plans and addressing access to energy in line with our relevant national commitment(s)'. Hence, while the commitment document and pledge is intent on supporting and producing plans and technical knowledge to respond to climate change challenges, the commitment is presented in a vacuum, removed from the tension and conflict that climate adaptation planning and energy access decisions can and do produce (Bulkeley & Tuts, 2013; Gore, 2017; Habtezion et al., 2015; Power et al., 2016). Critical concerns about climate justice at the urban scale are also glossed over at a time when the multiplicity of actors that are and will be involved in responding to climate change in African cities inevitably raises 'questions about who stands to gain and lose through such processes' (Bulkeley et al., 2014: 32).

There is a significant difference between coping with climate impacts and formally adapting, which entails long-term institutionalized responses (Bulkeley & Tuts, 2013). But to institutionalize adaptation requires a critical understanding of the political, ecological, and social context in each setting. The IPCC argues that well-governed cities, with universal provision of infrastructure and services, have a strong basis for climate resilience (Revi et al., 2014: 539). This, however, is not the norm in African cities. Further, the IPCC also argues that local government capacity also depends on vertical and horizontal coordination with national governments, the private sector and civil society. In most African cities, neither of these conditions – universal infrastructure access and strong vertical and horizontal coordination – exist. Infrastructure and services are poor and nowhere near universal; city and national governments conflict regularly due to cities being central sites of oppositional politics; and, local government fiscal and technical capacity is also often weak. So, while many African cities are pleased to make commitments to take climate mitigation and adaptation action, very few well documented cases of institutionalized adaptation exist.

One notable example where this has taken place is in Durban, South Africa (Carmin et al., 2012; D. Roberts, 2009). Durban's progress in institutionalizing climate adaptation planning was a result of endogenous leadership by non-elected officials – bureaucrats. Further, while the city did benefit from exogenous financial support for their climate activities, the city did not conform to national, regional or global norms but acted independently, engaging directly with the public and other municipal departments as a priority: 'This approach led adaptation to have rapid and widespread support' (Carmin et al., 2012: 29). While South Africa's financial and technical capacity far surpasses most other African countries and cities, the lessons from Durban resonate in other urban environmental governance fields in African cities. Cities can and do exercise agency, in spite of the complex and conflictual architecture that they are imbedded. This agency, like in Durban, is rooted in endogenous, long-term collaboration between unelected government officials and civil society. In the following section, the chapter compares climate adaptation with urban agricultural policy development in East Africa. The lessons resonate with the experience in Durban, and highlight the importance of independent, endogenous leadership, and where civil society engagement and leadership are central.

Architecture and Agency in African Cities: Comparing Climate Change and Urban Agriculture

The current and anticipated impacts of climate change on African cities are very well documented and have been for years (see Niang et al., 2014: 1224–1225); summaries emphasize the 'extremely high levels of vulnerability among the continent's large and rapidly growing urban poor populations' (Revi et al., 2014: 552). One of the most dominant risks to cities comes from flooding; flooding is pervasive in coastal cities and towns, and is extremely dangerous to vulnerable populations living in unplanned areas with poor drainage and precarious infrastructure (Douglas et al., 2009; Huq et al., 2007; Serdeczny et al., 2017). These concerns are accentuated by the potential for rainfall events to be highly concentrated in small geographic areas and due to the overall increase in the intensity of rainfall events, which are compounded by poor or non-existent infrastructure systems able to withstand high volumes of water (Revi et al., 2014: 556). With flooding comes the potential for the destruction of the central livelihood base for humans – their homes – and the spread of disease due to stagnant water, transportation cut offs, and food price volatility. These latter phenomena have been the subject of a high level of research and speculation, owing to the challenge of changing food production practices due to uncertain and complex weather patterns (Pinstrup-Andersen & Watson, 2011; Serdeczny et al., 2017; Toulmin, 2009; Yanda & Mubaya, 2011) and, more alarmingly, the potential for political instability and conflict due to inadequate food supplies and food price volatility.

A direct, causal connection among food price increases, urban protest, and political instability is not easy to discern, but the correlation is strong (Bellemare, 2015; Cohen & Garrett, 2010). Urban residents in African cities, like in all cities, rely predominantly on purchased food (Crush, Frayne & Pendleton, 2012). But the propensity for low-, middle- and high-income African urban residents to produce their own food for consumption and sale – to engage in urban and peri-urban agriculture – particularly during times of political and social unrest, is very well documented (Lee-Smith, 2010; Maxwell, 1999; Memon & Lee-Smith, 1993; Mougeot, 2005; Prain, Lee-Smith & Karanja, 2010). This trend globally and in Africa has not waned over time (Gore, 2018). Beyond the livelihood and nutrition benefits from the consumption of urban food production, there is a great deal of international support for urban food production as well, particularly in relation to climate change.

Global norms articulating the right to food and freedom from hunger predate any global climate norms (Gore, 2008). There are also recent global experiments in non-state orchestration platforms for urban agriculture, such as the Milan Urban Food Policy Pact. Signed by more than 165 cities around the world, representing more than 450 million urban inhabitants, signatories commit to 'develop sustainable food systems that are inclusive, resilient, safe and diverse', while seeking coherence between municipal, national, regional and international policies and processes, among other commitments (Milan Urban Food Policy Pact, 2018). The Pact also directly articulates the relationship among cities, climate change, and food systems. Organizations such as the RUAF Foundation also play a critical global role in promoting sustainable urban agriculture and food systems, and promoting the relationship between urban food production and climate change (RUAF Foundation, 2015).

Like international climate treaties, UN commitments and norms around food security produce a formal, complex global architecture for food access in cities. Further, like the Covenant of Mayors and other subnational climate networks, non-state orchestration platforms for urban agriculture articulate admirable norms, but ones that are also deeply political. Seeking policy coherence between African cities and national governments is enviable but deeply problematic. This is due to the fact that African national governments are notorious for weakly supporting cities or being openly antagonistic to them (Parnell & Simon, 2014), particularly when local governments promote urban agriculture (Gore, 2018). Hence, from the perspective of the global and national architecture in which climate action and urban agriculture are promoted in African cities, there are many parallels. African cities promote and engage directly with formal rules, authorities, and norms, while simultaneously joining subnational networks and forums that attempt to orchestrate local action. In both contexts – climate action and urban agriculture – these norms and forums promote coordination between international through to local authorities, but with little articulation of past conflict between national and local authorities in African states, and the conflict that ensues when the promotion of climate adaptation or the denial of urban agriculture ignores national–local tensions and/or does not engage directly with vulnerable urban residents. Indeed, when summarizing the lessons from an extensive review of climate adaptation, Satterthwaite, Dodman and Bicknell (2009) argue that adaptation needs to be locally driven and that local governments must be able and willing to work with inhabitants most at risk (p. 362). Put differently, agency – the capacity to influence and alter the authority and structures that influence outcomes – is expected to be shared and diffuse, and the result of prolonged or purposeful collaboration between city-based actors.

Recent advances in support for urban agriculture in the city of Nairobi, Kenya helps illustrate how collaborative processes between a diversity of indigenous urban-based actors produce stronger support for urban agriculture in the city. This form of 'bottom-up agency' (see Chapter 1 by van der Heijden, Bulkeley & Certomà in this book) contrasts with the cities of Dar es Salaam, Tanzania and Kampala, Uganda. In each setting, and compared to most other cities and states in Africa, the national and urban rules for urban agriculture in these East African cities were supportive. But it was in the city of Nairobi, Kenya where agency was most pronounced and most effective at altering the structures and authorities – the architecture – limiting the potential of urban agriculture.

Bottom-Up Agency in Urban Agriculture: Lessons from Nairobi, Kenya, Kampala, Uganda, and Dar es Salaam, Tanzania

There are several reasons to consider Nairobi, Kampala and Dar es Salaam together. All are the largest cities in their respective countries and all are experiencing very high population growth rates, well above the averages of other African states (Gore, 2018; UN-Habitat, 2008), with complementary problems in the provision of adequate housing and services. While their vulnerability to climate change varies owing to their respective qualities of infrastructure, geography, weather patterns, and settlement patterns, each has experienced, is experiencing, and will experience critical challenges relating to flooding, heat and human health as a result of climatic change (Government of Kenya, 2016; Lwasa, 2010b in Gore, 2015; START, 2011). In light of these concerns, it is noteworthy that one authority related to each city is a signatory to the Covenant of Mayors. The odd phrasing of the previous sentence is purposeful: While the Covenant notes each city as a signatory, only one of the cities is represented by a directly elected leader with control and authority over policy and finance for the entire city.

Kampala, Uganda is currently controlled by the national government. While the city has a popularly elected mayor and councillors, authority is vested in an Executive Director, appointed by the national government, and accountable to the national government (see Gore & Muwanga, 2014). The city of Dar es Salaam is made up of five independent municipalities. The Dar es Salaam City Council representatives are popularly elected, but the authority of the Council to coordinate actions between these municipalities is very weak (Gore, 2015). Further, the city of Dar es Salaam, officially, is also a national Region, which is overseen by a Regional Commissioner, appointed by the national government. Hence, the capacity of Dar es Salaam city (the region) to implement the goals of the Covenant is weak. Nairobi's signature to the Covenant is firmest with respect to being able to operationalize the commitments. Nairobi is led by a popularly elected Governor. This position arose following the passage of Kenya's new constitution in 2010, which established 47 subnational, county governments, including The County of Nairobi. These structural differences are important to bear in mind when considering how authority is exercised in each city for climate adaptation or urban agriculture. If collaboration between local governments and civil society is central to adaptation and responsiveness in environmental governance, the legitimacy of those in authority and the clarity of the nature of that authority are important.

With respect to urban agriculture, each city is recognized as having a long history of research on and support for urban food production (Lee-Smith, 2010, 2013; Prain et al., 2010). Agriculture in each country is also devolved to the subnational level, giving them authority to support agriculture as they determine. To various extents, each city and country has domestic or national policies or laws recognizing urban agriculture as a legitimate land use, even though the character of support and recognition have varied quite dramatically over time (Gore, 2008, 2018). But noting that agriculture is a legitimate land use in urban areas is not the same as producing updated bylaws or legislation, or civil society working collaboratively with local authorities to improve conditions for food production.

Around 2008, Kampala was expected to be the regional if not continental leader in urban agricultural support. It was one of the only cities in the subcontinent that had established bylaws guiding urban agricultural production. These bylaws were realized through a consultation process funded by a multilateral organization and involving a diverse range of local and national stakeholders (see Lee-Smith, 2010; Prain et al., 2010). Yet, shortly after the passage of the bylaws, the national government took over the city as tension between the national government and the city came to a head. Once the structure of the city authority and administration changed, so too did the relation among the city, farmers, and civil society. Following the national government's takeover in 2010, 'the connection between the new city government and civil society organizations, domestic academics, and farmers previously engaged in UA [urban agriculture] bylaw development was non-existent' (Gore, 2018: 177). No city staff previously involved in UA support remained after the city administration changed, and the interactions between civil society and the local state also ceased. While the city's support for UA remains, it is rebuilding its engagement with farmers and civil society slowly and is many years behind where it could have been.

In Dar es Salaam, the complexity of the city structure, its rapid pace of change, combined with deference to the national government and weak civil society (Dill, 2009), has meant very slow progress in supporting urban agriculture. Research by the author in Tanzania since 2013 has revealed a high level of uncertainty among civil society organizations, farmers and municipal agricultural officers about who should advocate for stronger urban agricultural support. Only one municipality (Kinondoni) in the city of Dar es Salaam was clearly advancing support for farmers, and like Durban, this was largely due to the leadership of a highly trained, energetic agricultural officer. In other municipalities in the city, agricultural officers said that they were very supportive of formalizing urban agriculture in their municipalities, but they were waiting and hoping that a long-delayed city Master Plan would eventually be released to formalize where urban agriculture could take place. (The Master Plan was more than two years behind in being released, and after being released, was widely criticized and dismissed, requiring further revision [Nachilongo, 2017]). Alternatively, agricultural officers suggested that advocacy for greater support had to come from civil society and farmers, not themselves. Conversely, farmers and civil society organizations expected municipal staff to take the lead in advocating for them as elected officials did not listen to them or support urban agriculture. The overall outcome in the city has been a continual waiting game, with the hope that the national government will eventually pronounce its strong support for urban agriculture in the city as a whole, giving a de facto 'green light' to increase support in each municipality.

Both Dar es Salaam and Kampala contrast with Nairobi. First, despite Kenya's history and legacy of political conflict, the structure of authority at the urban scale is clearer. Accountability for action or inaction in Nairobi is directly tied to the county administration. While this administration can certainly be influenced by innumerable sources, formal authority is not diffused between independent municipalities or between city and national authorities. Second, and most critically, civil society in Nairobi has played a critical and defining role in strengthening support for urban agriculture. One central organization, the Mazingira Institute, has been central to the strengthening of support for urban agriculture in the city (see Gore, 2018).

While other non-governmental organizations (NGOs) have been active in urban greening and agriculture in Kenya, Mazingira's capacity, longevity, and influence as an independent research and advocacy organization, particularly for issues relating to human rights, land rights, and the environment, is relatively unmatched. Most critically, the Institute's founders, Davinder Lamba and Diana Lee-Smith, have high political and policy acumen, and key staff members, particularly the late Kuria Gathuru, were exceptionally well engaged with citizens. Combined, these characteristics gave the organization a very high degree of effectiveness and influence.

Mazingira undertook seminal research on urban agriculture and land use in Kenya in the 1980s, which has served as reference point for other research since then. Further, in the early 2000s, as the political space in the country started to open, it engaged directly with critical staff in the Ministry of Agriculture to discuss urban agriculture and promote its importance. Over time, these relations strengthened, with trust and alliances developing. Not long after, in 2004, Mazingira created a forum for dialogue about urban agriculture in Nairobi (Gore, 2018). This forum was not created to try to produce a single output or policy goal, but to create a safe, collaborative, and critical space for formal authorities (bureaucrats), farmers, and advocates to engage in sustained dialogue. From these forums, agricultural training evolved, so too did smaller farmer and women-led support groups. But most critically with respect to advancing support for urban agriculture in the city was the fact that when the political space to support urban agriculture opened, a diverse network of interests, including those engaged directly in food production – the farmers – were ready and able to articulate their needs and interests in collaboration with civil society and the agricultural leaders in government. In Nairobi, critical alliances between unelected government agricultural leaders, civil society and citizens had been built from the bottom-up and over time: the collaborative forums 'embraced multiple actors, expertise from civil society, government and farmers, and built capacity, knowledge and relations between the various groups. The processes were deliberative, inclusive and multi-purposed' (Gore, 2018: 177).

From Nairobi, a key lesson is that agency developed slowly, from the bottom upwards. Farmers and civil society did not treat the city or national government as a whole, but recognized the importance of building trust and alliances with key members of government. The power of these alliances was realized through the quality of the interactions between diverse actors, which came to fruition when there was an opening in the political space and architecture of the city.

The experience in Nairobi resonates with the overall arguments for the conditions needed to successfully adapt to climate change – local governments need to engage directly with individuals most affected by climate change. But an additional observation is that the architecture of the local and national state can also play a critical role in facilitating when and how city alliances flourish and alter the responsiveness of the state. If the structure is closed, then the alliances may have to wait and continue their dialogue and collaboration until a window opens or can be forced open to enable the response needed and desired at the local scale.

Conclusion

One of the central deficits in research about global environmental change in Africa is how the quality of governance systems and processes relate to the production of policy, programmes and actions that meet human and ecological imperatives (Habtezion et al., 2015: 202). Understanding the nexus between democracy and environmental governance in African cities requires a historical and ongoing understanding of the architecture of city–national authority, and how agency has been exercised within these structures. Civil society groups understand well that 'The making and realization of policy is a game of power and conflicting interests' in African cities (Lee-Smith, 2010: 497). The challenge that remains is how to turn this game into a common practice for effective responsiveness to critical human and ecological challenges. In order to achieve these goals, the complexity of the relationship between architecture and agency in African states needs to be illuminated. This chapter sought to emphasize this complexity and to illuminate that the processes that are prescribed for achieving progress in environmental governance are being realized in some African cities, albeit slowly.

As Habtezion et al. argue (2015: 202–203), the large gap in analysis between the character of state and city democracy and ecological and human welfare outcomes in Africa needs to be rectified. There remains little empirical evidence about African city climate action or adaptation, particularly that focuses on the politics and conflicts imbedded in these processes. This is an empirical challenge for research on municipal governments globally, given the number of potential cases that exist. There is a tendency for researchers to have to choose between a high number of cases in order to generalize or to theorize through comparing several cases. Both approaches have value, but research attention and support must be given to studies that deeply probe the specific dynamics between architecture and agency in climate and environmental governance. Deep qualitative analysis will not necessarily allow for generalizability in the short term. But if it is well recognized that climate adaptation planning and policy can exacerbate political tensions in African cities, then it is necessary to understand deeply how agency is exercised and how authority is articulated in these settings.

This chapter has revealed that the relationship between architecture and agency, and the character of the political space in which these pillars of governance intersect, are critical for understanding how responsive policies and programmes can be realized in African cities. In many cases, international orchestration platforms can provide guidance, inspiration and much needed attention for African cities. But based on limited empirical findings to date, the character and quality of the political space in which environmental policy materializes in African cities, especially how citizens, civil society and local governments engage with one another to discuss and deliberate over responses to city needs, seem most critical for producing the urban transformations so desired.

References

Atela, J. O., Quinn, C. H., Arhin, A. A., Duguma, L., & Mbeva, K. L. (2017). Exploring the agency of Africa in climate change negotiations: The case of REDD+. *International Environmental Agreements: Politics, Law and Economics*, 17(4): 463–482. https://doi.org/10.1007/s10784-016–9329-6

Bellemare, M. F. (2015). Rising food prices, food price volatility, and social unrest. *American Journal of Agricultural Economics*, 97(1): 1–21. https://doi.org/10.1093/ajae/aau038

Betsill, M. M., & Bulkeley, H. (2003). *Cities and Climate Change: Urban Sustainability and Global Environmetnal Governance*. New York: Routledge. www.routledge.com/Cities-and-Climate-Change/Betsill-Bulkeley/p/book/9780415273794

Betsill, M. M., & Bulkeley, H. (2004). Transnational networks and global environmental governance: The cities for climate protection program. *International Studies Quarterly*, 48(2): 471–493. https://doi.org/10.1111/j.0020–8833.2004.00310.x

Bulkeley, H. (2010). Governing climate change in the city. *Urbanization and Global Environmental Change Viewpoints*, 4: 15–19.

Bulkeley, H., Carmin, J., Castán Broto, V., Edwards, G. A. S., & Fuller, S. (2013). Climate justice and global cities: Mapping the emerging discourses. *Global Environmental Change*, 23(5): 914–925. https://doi.org/10.1016/j.gloenvcha.2013.05.010

Bulkeley, H., Edwards, G. A. S., & Fuller, S. (2014). Contesting climate justice in the city: Examining politics and practice in urban climate change experiments. *Global Environmental Change*, 25: 31–40. https://doi.org/10.1016/j.gloenvcha.2014.01.009

Bulkeley, H., & Tuts, R. (2013). Understanding urban vulnerability, adaptation and resilience in the context of climate change. *Local Environment*, 18(6): 646–662. https://doi.org/10.1080/13549839.2013.788479

Carmin, J., Anguelovski, I., & Roberts, D. (2012). Urban climate adaptation in the Global South: Planning in an emerging policy domain. *Journal of Planning Education and Research*, 32(1): 18–32. https://doi.org/10.1177/0739456X11430951

Cohen, M. J., & Garrett, J. L. (2010). The food price crisis and urban food (in)security. *Environment and Urbanization*, 22(2): 467–482. https://doi.org/10.1177/0956247810380375

CoMSSA. (2018). About the Initiative. http://comssa.org/about/ (accessed 28 April 2018).

Crush, J., Frayne, B., & Pendleton, W. (2012). The crisis of food insecurity in African cities. *Journal of Hunger & Environmental Nutrition*, 7(2–3): 271–292. https://doi.org/10.1080/19320248.2012.702448

de Waal, A. (2018). *Mass Starvation: The History and Future of Famine*. Hoboken, NJ: John Wiley & Sons.

Dill, B. (2009). The paradoxes of community-based participation in Dar es Salaam. *Development and Change*, 40(4): 717–743. https://doi.org/10.1111/j.1467–7660.2009.01569.x

Douglas, I., Alam, K., Maghenda, M., McDonnell, Y., McLean, L., & Campbell, J. (2009). Unjust waters: Climate change, flooding, and the urban poor in Africa. In *Adapting Cities to Climate Change: Understanding and Addressing the Development Challenges*. London: Earthscan, 201–218.

FAO (Food and Agriculture Organization of the United Nations). (2011). *The State of Food Insecurity in the World*. Rome: FAO.

Global Carbon Atlas. (2018). Global carbon atlas. www.globalcarbonatlas.org/en/content/welcome-carbon-atlas (accessed 28 April 2018).

Global Covenant of Mayors. (2018). Commit. www.globalcovenantofmayors.org/participate/ (accessed 28 April 2018).

Goodfellow, T., & Titeca, K. (2012). Presidential intervention and the changing 'politics of survival' in Kampala's informal economy. *Cities*, 29(4): 264–270.

Gore, C. D. (2008). Healthy urban food production and local government. In Donald Cole, Diana Lee-Smith, & George Nasinyama (eds.), *Healthy City Harvests: Generating Evidence to Guide Policy on Urban Agriculture*. Lima, Peru: CIP/Urban Harvest and Makerere University Press, 49–65.

Gore, C. D. (2010). The limits and opportunities of networks: Municipalities and Canadian climate change policy. *Review of Policy Research*, 27(1): 27–46. https://doi.org/10.1111/j.1541–1338.2009.00425.x

Gore, C. D. (2015). Climate change adaptation and African cities: Understanding the impact of government and governance on future action. In C. Johnson, N. Toly, & H. Schroeder (eds.), *The Urban Climate Challenge: Rethinking the Role of Cities in the Global Climate Regime*. Abingdon, Oxon: Routledge, 205–223.

Gore, C. D. (2017). *Electricity in Africa: The Politics of Transformation in Uganda*. Woodbridge, Suffolk: James Currey.

Gore, C. D. (2018). How African cities lead: Urban policy innovation and agriculture in Kampala and Nairobi. *World Development*, 108: 169–180. https://doi.org/10.1016/j.worlddev.2018.03.011

Gore, C. D., Brass, J. N., Baldwin, E., & MacLean, L. M. (2018). Political autonomy and resistance in electricity sector liberalization in Africa. *World Development*. https://doi.org/10.1016/j.worlddev.2018.03.003

Gore, C. D., & Muwanga, N. K. (2014). Decentralization is dead, long live decentralization! Capital city reform and political rights in Kampala, Uganda. *International Journal of Urban and Regional Research*, 38(6): 2201–2216. https://doi.org/10.1111/1468–2427.12012

Government of Kenya. (2016). Kenya National Adaptation Plan: 2015–2030.

Habtezion, S., Adelekan, I., Aiyede, E., et al. (2015). Earth system governance in Africa: Knowledge and capacity needs. *Current Opinion in Environmental Sustainability*, 14: 198–205.

Huq, S., Kovats, S., Reid, H., & Satterthwaite, D. (2007). Editorial: Reducing risks to cities from disasters and climate change. *Environment and Urbanization*, 19(1): 3–15. https://doi.org/10.1177/0956247807078058

Kern, K., & Bulkeley, H. (2009). Cities, Europeanization and multi-level governance: Governing climate change through transnational municipal networks. *JCMS: Journal of Common Market Studies*, 47(2): 309–332. https://doi.org/10.1111/j.1468–5965.2009.00806.x

Kithiia, J. (2011). Climate change risk responses in East African cities: Need, barriers and opportunities. *Current Opinion in Environmental Sustainability*, 3: 176–180.

Klaus, K., & Paller, J. W. (2017). Defending the city, defending votes: Campaign strategies in urban Ghana. *The Journal of Modern African Studies*, 55(4): 681–708. https://doi.org/10.1017/S0022278X17000453

Krause, R. M. (2012). An assessment of the impact that participation in local climate networks has on cities' implementation of climate, energy, and transportation policies. *Review of Policy Research*, 29(5): 585–604. https://doi.org/10.1111/j.1541-1338.2012.00582.x

Lee-Smith, D. (2010). Cities feeding people: An update on urban agriculture in equatorial Africa. *Environment and Urbanization*, 22(2): 483–499. https://doi.org/10.1177/0956247810377383

Lee-Smith, D. (2013). Which way for UPA in Africa? *City*, 17(1): 69–84. https://doi.org/10.1080/13604813.2012.754177

MacLean, L. M., Gore, C. D., Brass, J. N., & Baldwin, E. (2016). The politics of state-building and access to electricity provision in Ghana and Uganda. *Journal of African Political Economy & Development*, *1*(1). https://afec-japed.com/expectations-of-power-the-politics-of-state-building-and-access-to-electricity-provision-in-ghana-and-uganda/

Maxwell, D. (1999). The political economy of urban food security in Sub-Saharan Africa. *World Development*, 27(11): 1939–1953.

Memon, P. A., & Lee-Smith, D. (1993). Urban agriculture in Kenya. *Canadian Journal of African Studies/Revue Canadienne Des Études Africaines*, 27(1): 25. https://doi.org/10.2307/485438

Mhagama, H. (2016, January 29). Tanzania: Dar es Salaam has five districts as Ubungo, Kigamboni added to it. *Tanzania Daily News (Dar Es Salaam)*. http://allafrica.com/stories/201601291386.html

Milan Urban Food Policy Pact. (2018). Milan Urban Food Policy Pact text. www.milanurbanfoodpolicypact.org/text/ (accessed 28 April 2018).

Mougeot, L. J. A., ed. (2005). *Agropolis. The Social, Political and Environmental Dimensions of Urban Agriculture*. London: Earthscan.

Mwita, I. (2016, September 13). C40: Lord Mayor of Dar es Salaam, Isaya Mwita, spells out his climate change priorities. www.c40.org/blog_posts/lord-mayor-of-dar-es-salaam-isaya-mwita-spells-out-his-climate-change-priorities (accessed 12 February 2018).

Myers, P. G. (2011). *African Cities: Alternative Visions of Urban Theory and Practice*. London: Zed Books.

Nachilongo, H. (2017, September 12). New master plan crucial for turning Dar into sustainable city. www.thecitizen.co.tz/News/New-master-plan-crucial-for-turning-Dar-into-sustainable-city/1840340–4091494-tbuneiz/index.html (accessed 30 April 2018).

Newell, P., & Bulkeley, H. (2016). Landscape for change? International climate policy and energy transitions: Evidence from sub-Saharan Africa. *Climate Policy*, 1–14. https://doi.org/10.1080/14693062.2016.1173003

Newell, P., & Mulvaney, D. (2013). The political economy of the 'just transition'. *The Geographical Journal*, 179(2): 132–140. https://doi.org/10.1111/geoj.12008

Niang, I., Ruppel, O. C., Abdrabo, M. A., et al. (2014). Africa. In *Climate Change 2014: Impacts, Adaptation, and Vulnerability. Part B: Regional Aspects*. Contribution of Working Group II to the Fifth Assessment Report of the Intergovernmental Panel on Climate Change [Barros, V. R., C. B. Field, D. J. Dokken, M. D. Mastrandrea, K. J. Mach, T. E. Bilir, M. Chatterjee, K. L. Ebi, Y. O. Estrada, R. C. Genova, B. Girma, E.

S. Kissel, A. N. Levy, S. MacCracken, P. R. Mastrandrea, and L. L. White (eds.)]. Cambridge University Press, Cambridge, United Kingdom and New York, NY, USA, 1199–1265.

Okanya, A. (2016, May 8). Musisi shares KCCA's experience in combating climate change. www.newvision.co.ug/new_vision/news/1424092/musisi-shares-kccas-experince-combating-climate-change (accessed 12 February 2018).

Paller, J. W. (2014). Informal institutions and personal rule in urban Ghana. *African Studies Review*, 57(3): 123–142. https://doi.org/10.1017/asr.2014.95

Paller, J. W. (2015). Informal networks and access to power to obtain housing in urban slums in Ghana. *Africa Today*, 62(1): 30–55.

Parnell, S., & Simon, D. (2014). National urbanisation urban strategies: Necessary but absent policy instruments in Africa. In E. Pieterse & S. Parnell (eds.). *Africa's Urban Revolution*. London: Zed Books.

Peters, G. P., Minx, J. C., Weber, C. L., & Edenhofer, O. (2011). Growth in emission transfers via international trade from 1990 to 2008. *Proceedings of the National Academy of Sciences of the USA*, 108(21): 8903–8908. https://doi.org/10.1073/pnas.1006388108

Pieterse, D. E., & Parnell, S. (2014). *Africa's Urban Revolution*. London: Zed Books.

Pinstrup-Andersen, P., & Watson II, D. D. (2011). *Food Policy for Developing Countries*. Ithaca, N: Cornell University Press.

Power, M., Newell, P., Baker, L., Bulkeley, H., Kirshner, J., & Smith, A. (2016). The political economy of energy transitions in Mozambique and South Africa: The role of the rising powers. *Energy Research & Social Science*, 17: 10–19. https://doi.org/10.1016/j.erss.2016.03.007

Prain, G., Lee-Smith, D., & Karanja, N., eds. (2010). *African Urban Harvest*. New York: Springer Science+Business Media. https://doi.org/10.1007/978-1-4419-6250-8

Resnick, D. (2012). Opposition parties and the urban poor in African democracies. *Comparative Political Studies*, 45(11): 1351–1378. https://doi.org/10.1177/0010414012437166

Revi, A., Satterthwaite, D., Aragón-Durand, F., et al. (2014). Towards transformative adaptation in cities: The IPCC's Fifth Assessment. *Environment and Urbanization*, 26(1): 11–28. https://doi.org/10.1177/0956247814523539

Roberts, D. (2009). Thinking globally, acting locally: Institutionalizing climate change at the local level in Durban, South Africa. In D. Dodman, J. Bicknell, & D. Satterthwaite (eds.), *Adapting Cities to Climate Change: Understanding and Addressing the Development Challenges*. London: Earthscan, 253–270.

Roberts, J. T., & Parks, B. C. (2007). *A Climate of Injustice*. Cambridge, MA: MIT Press. https://mitpress.mit.edu/books/climate-injustice

Robinson, P., & Gore, C. (2015). Municipal climate reporting: Gaps in monitoring and implications for governance and action. *Environment and Planning C: Government and Policy*, 33(5): 1058–1075. https://doi.org/10.1177/0263774X15605940

Roger, C., & Belliethathan, S. (2016). Africa in the global climate change negotiations. *International Environmental Agreements: Politics, Law and Economics*, 16(1): 91–108.

RUAF Foundation. (2015, January 30). Policy brief: Urban agriculture as a climate change strategy. www.ruaf.org/publications/policy-brief-urban-agriculture-climate-change-strategy-2015 (accessed 7 February 2018).

Sattherthwaite, D., Dodman, D., & Bicknell, J. (2009). Conclusions: Local development and adaptation. In J. Bicknell, D. Dodman, & D. Sattherthwaite (eds.), *Adapting Cities to Climate Change: Understanding and Addressing the Development Challenges*. London: Earthscan.

Serdeczny, O., Adams, S., Baarsch, F., et al. (2017). Climate change impacts in Sub-Saharan Africa: From physical changes to their social repercussions. *Regional Environmental Change*, 17(6): 1585–1600. https://doi.org/10.1007/s10113-015-0910-2

Sovacool, B. K., & Dworkin, M. H. (2014, September). *Global Energy Justice: Problems, Principles, and Practices*. Cambridge: Cambridge University Press. https://doi.org/10.1017/CBO9781107323605

START. (2011). Urban poverty and climate change in Dar es Salaam, Tanzania: A case study – START. https://start.org/publication/urban-poverty-and-climate-change-in-dar-es-salaam-tanzania-a-case-study/ (accessed 28 April 2018).

Stren, R. (2014). Urban service delivery in Africa and the role of international assistance. *Development Policy Review*, 32(s1): s19–s37. https://doi.org/10.1111/dpr.12067

Toulmin, C. (2009). *Climate Change in Africa*. London: Zed Books. http://africanarguments.org/climate-change-in-africa/

UN-Habitat. (2008). *The State of African Cities*. Nairobi, Kenya: UN-Habitat.

van der Ven, H., Bernstein, S., & Hoffmann, M. (2017). Valuing the contributions of nonstate and subnational actors to climate governance. *Global Environmental Politics*, 17(1): 1–20.

Yanda, P. Z., & Mubaya, C. P. (2011). *Managing a Changing Climate in Africa. Local Level Vulnerabilities and Adaptation Experiences*. Dar es Salaam, Tanzania: Mkuki na Nyota Publishers.

12

The Effects of Transnational Municipal Networks on Urban Climate Politics in the Global South

FEE STEHLE, CHRIS HÖHNE, THOMAS HICKMANN, AND MARKUS LEDERER

Introduction

Transnational municipal networks (TMNs) have emerged as important actors in the global response to climate change.[1] Environmentalists both within and outside governments have placed great expectations and hopes on these networks (Betsill & Bulkeley, 2007; Valente de Macedo, Setzer, & Rei, 2016). Through their novel form of agency, TMNs are expected to provide cities with a form of 'extra-legem' empowerment in their responses to climate change (Chapter 1 by van der Heijden, Bulkeley, & Certomà in this book). This empowerment is expected to work by linking the international with the local level, which will eventually allow cities to break out of national positions on climate change, overcome local obstacles to the mitigation of climate change, and initiate adaptation policies (Acuto, 2013).

Numerous scholars have recently contributed to this field of research. In particular, various authors have shed light on the forms of agency that TMNs exert, such as strengthening local capacity, the diffusion and exchange of best practices, and initiating and supporting climate projects and policies (Betsill & Bulkeley, 2004; Gordon & Johnson, 2017; Hickmann, 2016; Román, 2010). The literature has shown that TMNs can empower city governments and urban climate governance through enhancing policy learning, jump-starting processes of sustainable transformation and potentially bridging the lack of regulatory environments (Lee & Van de Meene, 2012; Toly, 2008). However, only a few studies have examined the extent to which subnational climate actions transform into continuous policies and stable organizational structures

[1] This book chapter is based on a research project entitled 'Carbon Governance Arrangements and the Nation-State: The Reconfiguration of Public Authority in Developing Countries', funded by the German Research Foundation (DFG); Reference Numbers: FU 274/11-1 and LE 2644/4-1; Project Number: 270088441. We thank the DFG for providing funding, all interviewees for sharing their insights, Harald Fuhr for valuable discussions and comments on this chapter, and the editors for very helpful comments und support.

at the city level (Anguelovski & Carmin, 2011). In addition, while a number of studies have compared countries in the Global North in this regard (Fisher, 2013; Selin & VanDeveer, 2012), little research has been carried out in countries of the Global South (for an exception, see Hickmann et al., 2017). Finally, despite the burgeoning literature on TMNs over the past two decades, we see a knowledge gap with regard to the effects of TMNs on urban climate governance.

We are therefore interested in the effects of TMNs on the empowerment of urban climate politics within national jurisdictions, what leads to lasting policy changes at the city level, and which forms of frictions might arise from their interplay with domestic contexts, questions that are also connected to the overall interest of the book. In this chapter, we thus ask the question of how and to what extent TMNs foster the institutionalization of urban climate politics in the Global South. We follow Streeck and Thelen in defining institutions as 'formalized rules that may be enforced by calling upon a third party', which can become manifest in policies and organizations (2005: 10). Focusing on two prominent networks, the C40 Cities Climate Leadership Group (C40) and ICLEI – Local Governments for Sustainability (formerly named International Council for Local Environmental Initiatives), we investigate their effects in member cities of four emerging economies (i.e. Brazil, India, Indonesia, and South Africa) in the period from 2005 to 2016. While cities in Brazil and South Africa generally exhibit an advanced degree of action on climate change, most cities in India and Indonesia pursue climate activities less ambitiously.

Against this backdrop, we analyse whether TMNs have made a difference in this variance and provide some explanations of why these different patterns have emerged: First, we differentiate between the modes of action the networks choose in reaching out to city governments. In a second step, we analyse whether their activities have inspired changes in the organizational setting and in terms of policies that could lead towards a gradual institutionalization of urban climate politics, or whether scattered pilot projects and insulated interventions remain once-off activities. We eventually examine enabling and constraining factors in the respective political, administrative, and economic systems, which considerably determine the extent to which TMNs can promote climate action in cities both at national and subnational levels of government.

TMNs and the Institutionalization of Urban Climate Politics

How and to what extent do TMNs incentivize the institutionalization of climate politics in cities of the Global South? To address this research question, we first review the literature on transnational climate governance with regard to the modes in which TMNs interact with cities. To understand the dynamics of organizational and policy change and the domestic factors shaping such processes, we subsequently discuss insights from the field of public policy. We do this in a two-step approach by first outlining how to evaluate whether and to what degree TMNs had an effect on cities' climate policies and organizations. Second, we draw on insights of the public policy literature in order to identify enabling or constraining factors in the national context that can explain to what extent TMNs influence policy and organizational change in cities.

The consequences of the emergence of TMNs have been conceptualized as a 'reconfiguration of political authority across multiple levels and between public and private actors' (Bulkeley, 2010: 231). Studies on ICLEI's Cities Climate Protection program have demonstrated how the program incentivized the improvement of cities' previous actions on climate change (Kousky & Schneider, 2003). Five main factors that shape the effect of the program on member cities have been identified: climate action was more likely if a city government had (1) committed individuals, (2) regulatory power in key areas, such as energy and transportation, (3) sufficient financial resources, (4) co-benefits of climate action, and (5) political will (Bulkeley & Betsill, 2003). Consequently, TMNs are expected to contribute to the emergence of new activities and ultimately to the institutionalization of urban climate politics in cities, a finding that has underlined the necessity of further theoretical clarification and empirical investigation on the precise impact of TMNs in this process (Bulkeley et al., 2003).

Despite the recognition of this decisive role of TMNs, we still lack a thorough understanding of the ways in which TMNs induce specific climate policy outputs and outcomes in their member cities, especially in the Global South (Fünfgeld, 2015). A few large-N studies, which have addressed this research gap so far, find a positive correlation between a city's membership in a TMN and the adoption of climate policies (for an overview, see Rashidi & Patt, 2017). However, these studies lack a focus on the ways in which policies are transferred into their local context, and the processes by which policies are adopted. Moreover, most contributions studying the effect of TMNs on urban climate politics use the number of projects implemented as an indicator for a city's action on climate change, and focus on the role of policy entrepreneurs as agents of change in the local context (Chan et al., 2018). However, the number of projects alone does not indicate whether we can witness an overall policy or organizational shift. Furthermore, the multilevel nature of urban climate politics requires taking into account intergovernmental relations when analysing processes of change. The interplay of TMNs with the respective local and national policy context is thus a crucial, and yet understudied research strand.

To trace the interaction between TMNs and cities, cases can be categorized based on Andonova et al. (2009), who establish three modes of TMN governance: information-sharing, capacity-building and implementation, and rule-setting. We focus on the engagement of TMNs in urban climate politics and adapt this categorization by distinguishing five modes of TMN governance: (1) generation of information and monitoring; (2) support and capacity building to projects, strategies, and policies; (3) peer-to-peer learning; (4) engagement with political leadership and diplomatic involvement; and (5) brokering of finance. We thus seek to understand whether the involvement of cities in TMNs has an effect on the extent to which urban climate politics are institutionalized, and which type of change is occurring. Concepts of the public policy literature on policy and organizational change are a helpful tool for this undertaking. Organizational and policy change can be operationalized in a continuum that ranges from no change, over incremental change, to radical change (Capano, 2009).

For the context of urban climate governance, the initial stage of policy change entails the set-up of pilot projects or the set-up of greenhouse gas (GHG) inventories. A medium change implies the set-up of climate action plans and concrete GHG reduction targets, while the adoption of a climate policy or law can be considered to be a major change. Finally, a radical policy change is present when policies in sectors such as transport or energy include mitigation targets, or are adjusted accordingly (based on Lee & Koski, 2014). On the other hand, organizational change is small when a few people work on climate change issues while also fulfilling other tasks, whereas the set-up of a climate change unit in environmental and or sectoral departments is a medium change. Major changes include the additional establishment of functional cross-sectoral committees. Finally, the integration of climate change into strategic or planning units or within the core executive of a city (e.g. the mayor's office) depicts a radical organizational change (see e.g. Kern & Alber, 2009).

But how to explain such variation, and which factors enable or constrain long-term policy change? In searching for explanations of the extent to which external agency can bring about change in domestic contexts, scholars of international relations and public policy have extensively studied the influence and mechanisms of international and transnational organizations on policy-making in domestic bureaucracies (Busch, Jörgens, & Tews, 2005; Dolowitz & Marsh, 1996; Rose, 1993 for an overview, see also Lederer, 2018). They contend that external drivers and the extent of their impact on public-administrative structures and processes in nation-states vary extensively, and ascribe this variance to distinct policy contexts, consisting of historical patterns and domestic institutional environments (Dolowitz & Marsh, 2000).

Scholars recently have made a similar observation, arguing that transnational governance does not take place in a void, but depends to a large extent on economic, social, and political variables in the national context in which the subnational actors participating in transnational governance are embedded (Roger, Hale, & Andonova, 2017). Besides analysing the modes in which TMNs interact with local actors and the effects this causes, it is therefore of equal importance to take account of independent variables in the specific policy context (Howlett, Ramesh, & Perl, 2009; Peters & Pierre, 2006). Following Hall (1997) in assuming that interests, institutions, and ideas shape organizational and policy changes, and drawing from various strands of the (neo-institutional) literature on policy change, four main factors in the respective domestic policy contexts can be identified that have an enabling or constraining effect on the workings of TMNs.

First of all, tradition and path dependency can shape the set-up of nation-states and the distribution of authority (Carpenter, 2010; Lodge, 2012). This includes historical developments, questions of reputation, and the history of institutions, for example, the necessity of political centralization for emerging states. Historical institutionalists go even further by stating that existing and past arrangements of public policies cause path dependencies in current policy choices (Thoenig, 2012). As a consequence, they see radical policy changes as largely impossible, and consider changes to only take place over long-term periods through a gradual influence on institutions (Streeck & Thelen, 2005).

A second factor encompasses characteristics in the respective national political-administrative system. They include the federal or unitary nature of the political system and the degree of fiscal, administrative, and political decentralization. These characteristics shape the extent to which local governments can respond to climate change (Lodge, 2012), including the extent to which a municipality can locally collect and redistribute taxes, organize local government structures and to design its own policies and long-term strategies (Garschagen & Romero-Lankao, 2015). In addition, the mandate of a local government to provide transport, energy, and waste services shapes the extent to which it can induce GHG emission reductions. Last, the ability to access external loans shapes the extent to which local governments can finance costly sustainable transformations of infrastructure, such as metro-systems, clean bus fleets, or renewable energy generation.

Questions of political-economic structures can be identified as a third factor, including the dominance of certain economic interest groups and political departments and the degree to which they influence policy outputs and outcomes (Bulkeley, 2010; Lodge, 2012). Finally, political leadership and prioritization is a fourth factor, which can open or close a window of opportunity for external influence on domestic politics. Political preferences and leadership furthermore play a decisive role when it comes to the institutionalization of policies and organizational units at the local level (Kousky & Schneider, 2003; Roberts, 2008).

In the following section, we employ the analytical frame established in this review by focusing on two important TMNs, ICLEI and C40. While ICLEI has supported climate-related activities of cities and their administrations over a long period, C40 creates an exclusive club of the world's megacities. Both networks focus on abating emissions and on fostering the implementation of projects and action plans that set urban GHG emission reduction targets (Toly, 2008). The cities we analyse are located in countries that display an array of different political, administrative, and economic systems. This allows us to examine the relation between the influences a network can exert on climate politics of its member cities, and at the same time point out constraining or enabling political, administrative, and economic factors in the respective domestic context that shape the institutionalization of urban climate politics.

Enablers of Change? TMNs and Their Effects on Cities

In the following, we portray our empirical contribution to the research on the politics of urban climate futures. Opening with a comparison of the modes of engagement of the two networks, we trace the degree of policy and organizational change across the cases and present evidence on how and to what extent the involvement of cities in TMNs has supported or initiated the same. Finally, we identify empirical evidence for a set of factors in the respective domestic context that can hinder or facilitate the institutionalization of urban climate politics.

TMNs and Their Modes of Action

C40 and ICLEI have member cities in all of our four country cases. In this chapter, we focus only on some of them: in Indonesia, Jakarta is the only C40 city. Balikpapan and Bogor are among the active ICLEI member-cities, while Bandung is less involved. South Africa's three biggest cities Johannesburg, Cape Town, and Durban are active C40 and ICLEI members. In Brazil, São Paulo, Rio de Janeiro, Curitiba, and Salvador are members of both C40 and ICLEI. Belo Horizonte is only a member of ICLEI. Many Indian cities are members of ICLEI and some of them, such as Thane and Rajkot, have actively been engaged in ICLEI project activities. C40 has rather few activities in the country. While New Delhi and Mumbai have been longstanding members, C40 has classified them as temporarily inactive. Bangalore, Jaipur, Kolkata, and Chennai joined C40 only very recently, in

Table 12.1 *Modes of TMN action*

	ICLEI	C40
Generation of information	Consultants, online tools, tailored information on policies, best practices and techniques	Information and review on demand of policies, techniques and best practices
Support and capacity building	Tailored advocacy, consultants, training, policy writing and planning support	Tailored support to cities, networking
Peer-to-peer learning	Study tours, workshops, webinars, case studies, national networks	Study tours, workshops, webinars, best practice reports
Engaging political leadership	Through technical staff, executive heads, mayors	Through mayors and technical staff
Finance	Facilitating contacts and attracting donor projects for member cities	Depending on city, support to apply for grants or facilitating contacts

2015 and 2016. In Table 12.1, we provide an overview on the different modes of engagement. However, as we point out in the following, the intensity of the networks support strongly differs across countries and cities.

Generation of Information

ICLEI supported cities in establishing GHG inventories, mostly by hiring external consultants. Rio de Janeiro was able to improve an early inventory, and Durban and Cape Town established first GHG inventories for their municipal operations. São Paulo established its first complete inventory in 2005 and Belo Horizonte followed in 2009. ICLEI supported Rajkot, Thane, Bogor, and Balikpapan to establish inventories. In India, ICLEI also used online tools for sharing information and best practices to support cities in implementing national programs, such as the Smart Cities Mission.

C40 provides cities on demand with information about specific policies, techniques of measuring, and best practices through its internal research centre. Since 2014, the network also has encouraged member cities to report to the Global Protocol for Community-Scale Greenhouse Gas Emission Inventories (GPC), developed in collaboration with WRI and ICLEI. We find varying degrees of support to cities for reporting: For example, in Cape Town and Johannesburg, C40 provided training for staff members and in 2014 supported the establishment of Johannesburg's first inventory. In Brazil, only Rio and Salvador report to the GPC, while C40 currently works with Jakarta on adopting the GCP standards.

Support and Capacity Building to Projects, Strategies, and Policies

In India, ICLEI strives to align the national, state, and local level to support cities' actions. The network provides advocacy to the Indian government on the Smart City Mission and the Solar Cities Program. The latter was inspired and scaled up from previous ICLEI projects with participants such as Rajkot. ICLEI also implements projects funded by development agencies and supports cities in the participation of national programs. For example, Rajkot receives support through the hiring of consultants, or training of city staff.

Furthermore, ICLEI supported the development of climate action plans and tried to facilitate their integration into overarching development plans in Rajkot, Thane, Bogor, Balikpapan, and Belo Horizonte. In Sao Paulo, the network assisted in the process of including sectoral emission reduction targets in the policy based on the precedent inventory. In Belo Horizonte, ICLEI supported a program on voluntary carbon payments by local companies.

Initially, C40 focused on networking among cities only. However, from 2009 until 2012, its implementing partner, the Clinton Climate Foundation, seconded staff to city administrations, such as Jakarta and Sao Paulo, to support the implementation of projects (e.g. energy efficiency). Since 2013, C40 has ended its staff assignments and instead opened continental offices as it realized that city specific support was needed. In Johannesburg, C40 reviewed the city's climate action plan and aims at facilitating the alignment of Jakarta's climate ambition level with the Paris Agreement. Moreover, C40 supported Jaipur to participate in the Indian Smart Cities Mission program. But in most member cities in India and Indonesia, it does not have the resources to provide tailored support.

Peer-to-Peer Learning

ICLEI fosters the exchange between cities in various ways. In a study tour, city staff from Bogor and Balikpapan went to Warsaw to learn from the local public transport system and to exchange solutions. In Brazil, together with the Konrad Adenauer Foundation, ICLEI supports the conference of environmental secretaries of the 27 largest cities, the CB27.

C40 supports implementation processes mostly within the frame of thematic networks. In this context, it organizes study tours, working sessions, and webinars for the exchange among cities. On transportation, Jakarta has engaged with Latin American cities on bus rapid transit systems and has learned from Istanbul. In various publications, C40 highlights cities' best practices, such as Delhi's waste to energy plant or the prioritization of sidewalks and biking in Chennai.

Engagement with Political Leadership and Diplomatic Involvement

ICLEI engages with technical level staff, commissioners, and mayors in all countries. Durban mayors became vice presidents of the network to enhance their political support to the local agenda on climate change, and the city and ICLEI closely collaborated in preparing the COP23 in 2011. In Latin America, ICLEI represents the Compact of Mayors and puts considerable effort into persuading cities to sign the commitment.

C40 engages with political leaders to foster local climate politics and external network activities. In Cape Town, it helped to increase the importance that the city mayor attributed to the local climate agenda. At the 2007 C40 conference in New York, São Paulo's mayor pronounced his commitment to a municipal climate policy with concrete reduction targets. In the run-up to various COPs, C40 engaged with city staff and mayors to present their cities' GHG emission reduction targets and projects. Since 2016, C40's steering committee (with Jakarta until recently) has been developing a city roadmap regarding the Paris Agreement.

The engagement with political leadership can also lead to conflicts. In Brazil and South Africa, ICLEI had their office within city governments, but moved outside municipal administrations after various struggles to cooperate. C40 has encountered similar difficulties: After a change of government in São Paulo, in 2013, the previously fruitful relationship between the network and the city stopped, and most projects halted.

Brokering of Finance

Across all countries, ICLEI offers facilitation of contacts to development partners through its own projects and for infrastructure climate funding. By inviting cities such as Balikpapan and Belo Horizonte to UNFCCC COP meetings, ICLEI provides them with a platform to present their plans to stakeholders, including financial institutions.

C40 has not brokered private sector funding but connected cities with donors for financial support. In Jakarta, C40 offered to broker private buyers of vacant lands to develop zero carbon projects. Bangalore, Kolkata, and Chennai made use of support offered by C40 to improve their creditworthiness. In Johannesburg, C40 helped the city to obtain a major loan from the French development agency (Agence Française de Développement [AFD]) and the Global Environment Facility (GEF) for its densification project Corridors of Freedom. In Curitiba, a project brokered through C40 could not take place because the city showed a lack of commitment.

Institutionalization of Urban Climate Politics

After demonstrating the different modes in which TMNs have interacted with cities in our country cases, in this section we trace the effects their actions had on actual policy and organizational change in the cities. We illustrate the different degrees of institutionalization of urban climate politics by tracking organizational and policy change over the period from 2005 to 2016 according to the continuum we introduced in the section 'TMNs and the Institutionalization of Urban Climate Politics'.

Organizational Change

In Brazil, the set-up of organizational units on climate change took a slow and diverse path, rather inspired by internal efforts and only partially supported by TMNs. Rio (2000), São Paulo (2003), and Belo Horizonte (2005) established climate units within their environmental secretariats. In Sao Paulo, diplomatic relations between the city's international relations secretariat and the City of London fostered climate related measures, and the city to be a founding member of C40 in 2005. In 2009, the city created an intersectoral committee, chaired by the planning secretariat, to coordinate the implementation of São Paulo's climate policy. Had not most activities been halted three years later, the case of São Paulo would be exemplary of radical organizational change. In Belo Horizonte, the affiliation with ICLEI encouraged the set-up of an intersectoral committee on climate change, which is less functional, but still fulfills criteria for a major change. Rio established a climate committee, which did not convene on a regular basis. The city thus depicts a case of medium organizational change. Curitiba represents a case of small organizational change, with only one staff member working permanently on climate change.

In South Africa, varying degrees of organizational change occurred, rather related to cities' differing preferences than to the influence of TMNs. Cape Town displays major organizational changes. The Energy and Climate Change Unit considerably expanded in members of staff and technical capacity over the past decade, and cross-sectoral coordinating bodies with a focus on mitigation and energy efficiency were created. Committed public officials and a local network of non-governmental organizations (NGOs) and academia mostly influenced these changes, while C40 only partially supported the process by training public officials. In Durban, we find evidence for medium organizational change, but no indication that it was inspired by TMNs. In reaction to electricity shortages, the city established an Energy Office with a mitigation unit in 2008 to improve energy efficiency measures. Johannesburg only shows small organizational changes. After the climate change officers' position was vacant for a long time now two city officials work on the topic. Intersectoral coordinating bodies do not exist, but the environmental department can theoretically influence service-providing entities under its supervision. C40 was involved in improving the unit's capacity.

In India, organizational changes are in the range of non-existent to small and are not, or not directly, influenced by TMNs. Some cities do not even employ an environmental engineer, and no city has created the post of climate change engineer. However, according to their own account, ICLEI has incentivized the hiring of renewable energy engineers. In the case of Rajkot, one ICLEI staff was placed in the city administration on a temporary basis. A couple of cities have an informal core group on climate change but involved staff members have other responsibilities in addition. For membership in C40, Delhi's Urban Development Department was the contact organization instead of the Environment Department, which is responsible for climate change among other environmental issues. Bangalore has no contact person for climate change issues at all.

In Indonesia, we find evidence for medium to major organizational changes, but also no indication that TMNs induced these changes. Across cities, staff members in Environment or Development Planning Agencies work on climate change. Bogor and Jakarta formed small climate change units within their Environment Agencies, representing medium organizational changes. Balikpapan and Bogor also established a coordinating team on climate change with staff from different agencies, which demonstrates a major change. However, most Indonesian cities have only slightly increased the staff and budget for climate change. Even though TMNs have not induced these organizational changes, we find indications that by providing capacity building, they have partially supported them.

Policy Change

Brazilian cities have undertaken their own efforts and led policy-making on climate change. The pioneer city was São Paulo, which passed its law on climate change in 2009 with a GHG reduction target of 30 per cent of 2005 levels by 2012. The city commissioned ICLEI and local actors from academia and civil society with the drafting of the policy. It stipulated the construction of landfill gas to energy stations and an increase of biofuel powered buses to achieve emission reductions. In 2016, São Paulo had neither reached the reduction targets nor integrated climate change in sectoral policies, which is why we attest a major rather than a radical policy change. In Rio, where C40 has supported the making of the 2011 climate change law, we also find major policy change. The city has implemented a Bus Rapid Transit System (BRT), but sectoral policies remain largely uninfluenced. ICLEI supported the discussions and drafting of the municipal law for reducing GHG emissions in Belo

Horizonte. Even though the city encountered problems in transforming the transportation profile, it still depicts a major policy change. Curitiba has no climate policy in place and thus demonstrates a case of small policy change.

In South Africa, we find evidence for major policy changes, but only in one case a TMN involvement. Cape Town set up the Integrated Metropolitan Environmental Policy (IMEP) as early as in 2001, and established a complementary set of mitigation strategies during the following 15 years. The city developed these major policy changes in collaboration with local NGOs and research institutions, rather than with support from TMNs. In 2015, Durban developed its Climate Change Strategy, which set the target to implement 40 per cent renewables by 2030. Lacking integration of mitigation targets into departmental agendas, it demonstrates a case of medium policy change. Finally, in Johannesburg we find evidence only for small policy change, even though C40 tried to induce the development of a climate change policy. The network supported the city in establishing its GHG inventory, and its research department reviewed the climate strategy, but congress has not approved it so far.

ICLEI has been instrumental in small policy changes such as setting up GHG inventories in Bogor, Balikpapan, Thane, and Rajkot. In Rajkot ICLEI supported pilot projects on solar energy and energy efficiency, and the city government scaled up an ICLEI initiative on LED street lighting. Most other small policy changes include sectoral pilot activities or business as usual actions that do not change the overall logic of the sector policy and were not caused by TMN influence. These include LED lamp installation and waste management in Balikpapan and Bogor, extension of green open space in Jakarta and Bogor, pilot projects on energy efficiency in public housing in Jakarta, fuel switching to compressed natural gas in buses in Jakarta and Bogor, as well as the procurement of electric buses in Delhi and Bangalore. Through the support of ICLEI, Rajkot developed a low-carbon development plan, representing a medium policy change.

ICLEI and C40 have also supported several activities by Indian and Indonesian cities which can be characterized as medium policy changes. Bogor, Thane, Rajkot, and Balikpapan have developed their low-carbon development plans with support of ICLEI. Bogor and Balikpapan integrated them in their mid-term development plans. Bogor, Balikpapan, and Rajkot also formulated GHG emission reduction targets. Jakarta independently developed its 2009 Provincial Climate Action Plan and formulated a GHG reduction targets as part of an ICLEI project in the run-up to the 2009 Climate Change Conference in Copenhagen on the request of C40. Furthermore, the city has independently integrated some climate considerations in the mid-term development plan and in the Spatial Plan for 2030.

Radical policy change has not occurred so far. However, while unsustainable development in Indonesia continues, radical changes might take place in the near future. For example, Jakarta aims to change its building code to reduce energy usage and GHG emissions by 30 per cent by 2030. The city is expanding its public transportation system by improving the BRT system and establishing a metro and light rail train network. While the planning process started long before Jakarta committed to climate change, the exchange with C40 working groups was beneficial. In turn, ICLEI's and C40's contribution to the development of public transportation systems in Bogor, Balikpapan, Bangalore, and Delhi was small to non-existent. Thus, for most of these sectoral developments, TMNs or even climate change considerations have not been instrumental. In contrast, C40 in some instances relabels Indian cities previous actions as climate relevant, for example, promoting Bangalore's comparably unsustainable waste management as very progressive.

Enabling or Constraining Factors in the Policy Context

Based on our previous exploration, we draw the preliminary conclusion that external actors such as TMNs can inspire policy and organizational change across cities in the Global South only to a small to medium extent. We find that an additional set of domestic factors specific for each nation state strongly defines whether such changes occur. We have structured these factors in the four main groups of (1) tradition and path dependency, (2) characteristics of the national political-administrative system, (3) the political-economic context, and (4) political leadership and prioritization. To illustrate to what extent these factors are enabling or constraining urban climate politics, we provide evidence for our argument with examples from the country cases.

Tradition and Path Dependency

Since the 1950s, India and Indonesia have had a long tradition of state-led development with centralized decision-making and multi-year planning. This has led to an overall top-down culture from political leaders to bureaucrats as well as from the national to the city level, preventing bottom-up initiatives and proactive engagement by cities and bureaucrats. While in the past, Indian and Indonesian cities have not been keen to look for inspiration in cities of other countries, TMNs were able to improve this situation.

In South Africa, past societal developments continue to shape the action of governmental actors on climate change to this day. The African National Congress (ANC) was the resistance movement under apartheid and has been the ruling party since the end of the Regime in 1994. A strong focus on pro-poor development provides legitimacy for the party's constituents, whereas environmental concerns are perceived as representing previous mechanism of exclusion and land expropriation by the then ruling white minority.

As a result of trade sanctions, South Africa relied on energy self-sufficiency under apartheid. Mining and energy companies were offered favourable deals to foster investment, and a state-owned entity controlled energy generation. These old networks of economic and political power in the sector continue to shape energy policy and have been the largest barrier for implementing renewable energies.

Characteristics of the National Political-Administrative Systems

Despite Brazil's federal system, decision-making in core sectors can still be highly centralized. Even though the constitution of 1986 grants municipal governments' strong autonomy, their competences remain constrained by a low degree of fiscal powers. Cities only receive a small share of 15 per cent of the total public revenues. In addition, the municipal budgets are bound to a large extent by salaries and conditional transfers on health and education. Only 10 per cent of the municipal budget remains for other expenses, climate change being one of them.

There are similar constraints for urban climate politics in the Indian federation. State governments and national government guidelines and programs shape the development of cities that lack powers, resources, and staff. The municipal commissioner, as the executive head of the city, is more powerful than the mayor, who has a rather ceremonial role. Municipal commissioners, who are state bureaucrats appointed by their respective state government, are more accountable to the state government. They can deliver changes when backed by their state government, mostly through the same ruling party, as it is the case in Rajkot/Gujarat concerning solar energy. National missions, such as the Jawaharlal Nehru National Urban Renewal Mission, the Smart Cities Mission, and the Atal Mission for Rejuvenation and Urban Transformation, guide cities' urban actions. As energy generation is a state function, cities have only limited room to manoeuver in determining their energy mix, for example by retrofitting municipal buildings. The only mandates cities possess are in the sectors of waste management and street lighting.

Even though a trend towards centralization is present in laws in 2004 and 2014, cities in Indonesia have been holding powers in many urban policy fields such as transport, waste, street lighting, and building codes since 1999. However, important issue areas such as energy production and distribution are highly centralized. In the sectors of transportation and waste, cities and provinces partly share responsibility. The resulting governance fragmentation is a major challenge. Cities lack financial resources, but benefit from national programs and funding, for example in a program to expand green open spaces in cities.

While the South African constitution grants autonomy to cities in a number of functions, a strong fiscal control of the national government limits their ability to implement larger transformational change. Furthermore, in the absence of a binding national legislation on climate change, urban climate action remains an unfunded municipal mandate. Municipalities can only marginally interfere in energy generation, centralized under the parastatal Eskom. Beyond that, South African metropolises shy away from introducing renewable energy even in legal grey areas, as they gain almost a third of their budget from surplus tariffs for the distribution of Eskom generated energy.

A lack of knowledge and capacity on climate change in city administrations is a further strong constraint for TMN actions. In India, cities heavily rely on consultants for preparing proposals for national programs (e.g. the Smart Cities Mission) or policies. Here, to implement best practices from abroad, ICLEI's staff or consultants have to accompany peer-to-peer learning in city networks. In addition, the continuous change of city staff complicates TMNs capacity-building activities across many cities, but especially in Indonesia, India, and Brazil.

Political-Economic Context

Brazil's national energy policy has counteracted municipal efforts to address increasing GHG emissions in the transport sector in the past years. After incentivizing the production of ethanol for more than four decades, the recent discovery of offshore oil off the coast of Rio de Janeiro changed preferences of important actors. Subsequently, the national government subsidized cars with diesel motors and lowered taxes on diesel, causing a significant increase of emissions in the transport sector. High subsidies on fuels in Indonesia have also resulted in a boost of individual traffic until they were reduced recently.

Clashes in the transition of local transport systems take place across all cities. In Brazilian cities, public concessions to private bus companies are set up to 30 years and encumber the transition to cleaner bus fleets. In addition, companies profit through reselling outdated models, and have less incentive to invest into sustainable and long-living vehicles. Furthermore, disruption in the employment opportunities in the informal transport business can hinder the extension of public transport systems. Protests took place in all South African cities, as well as in Delhi, Balikpapan, and Bogor.

Fossil fuel based energy generation as well as the lobbying by interest groups can also prevent cities from implementing mitigation actions. In South Africa, energy generation still relies for more than 90 per cent on coal. In the past, the parastatal Eskom has sought to maintain clientelistic structures and prevented an extensive inclusion of renewable energy into the countries energy mix. In contrast, favourable market and industry development has supported solar rooftop installations by Indian cities in municipal buildings. However, the strong lobbying of the coal industry and a still growing energy demand hamper the overall shift away from coal. In Indonesia, an increasing reliance on coal in the energy mix prevents cities from introducing renewable energy. Furthermore, high fossil fuel subsidies have undermined local efforts to increase energy efficiency in the past.

Political Leadership and Prioritization

Political leadership and their prioritization of climate change is a very important factor across all cities examined. In Indonesia, this is particularly evident, as there is a lack of bottom-up pressure by local communities to take climate actions. Mayors and commissioners in Bogor and Rajkot show interest in climate change and thus participate in many ICLEI projects. Under Governor Ahok, Jakarta has increasingly engaged on sustainable transportation and green building. The opportunity to shine on the global level at COPs also motivates mayors such as from Bogor to take actions.

Whereas we found strong evidence for the positive influence when a mayor supported climate action, we also found cases in which a change in leadership dampened or even reverted the climate agenda. In Brazil, this has been the case in São Paulo, where, after a fruitful period with a number of policies and sector-based emission reduction targets, the new mayor halted all activities in 2013 in exchange for a focus on pro-poor development. In Indonesia, the change of leadership has dampened climate actions in Jakarta. In India, changing municipal commissioners and state government leaders are much more important than changing mayors are. Political priorities across all cities often focus on problems in basic service provision or developmental aspects, which ICLEI tries to link with climate change.

In cases such as Cape Town, the presence of local champions in combination with strong links to the local epistemic community has had a highly positive influence on the development of planning and policy-making. However, even for cities with such favourable conditions, it is extremely difficult to overcome barriers in the political, administrative, and economic domestic context.

Conclusions

In this chapter, we explored the modes of engagements of TMNs with cities of the Global South, and the extent to which the involvement in TMNs had an effect on organizational and policy changes in these cities. We identified barriers and facilitators to their engagement in the political, economic, and administrative context of the respective country. Regarding the involvement of a city in a network, we find an interesting variance across all cities.

Based on our empirical findings, we draw three general conclusions. First, we identified some initial effects of the activities of TMNs on urban climate politics in our case study countries. In particular, there is evidence that TMNs, through their engagement in the dimensions portrayed earlier, can support at least some organizational and policy changes across many cities. However, their involvement alone is not sufficient to guarantee change, and, when singled out, can account for incremental changes, if any. This becomes obvious in Cape Town, which stands out in both organizational capacity and policy density on climate change. There, it was mainly larger local alliances between city officials, NGOs, academia, and donors in collaboration with TMNs who fostered change. In Brazilian metropolitan areas, TMNs have also supported changes, but depend on alliances with local epistemic communities and champions to a comparable extent.

In India and Indonesia, mostly ICLEI, and rarely C40 supported climate planning and learning in city governments. The low degree of political decentralization in Indian cities strongly restrains their involvement, and in some cases, has entirely prevented the occurrence of change. Project activities that depend on donor money often support changes, but they are temporary and mostly do not result in radical changes. Concerning the literature on urban climate governance, we thus conclude that TMNs can provide opportunities for policy learning, which often results in small policy changes, but not to a radical change towards the transformation of sectoral policies. Also, it seems important to note that even if policy changes are important to pave the way for institutionalizing urban climate politics, they can also remain empty symbolic gestures or window dressing.

Second, we did not find support for the claim that TMNs substantially revolutionized the way in which local climate policies are carried out or act as strong drivers for climate action. In a similar vein, we find that a TMN's intervention alone can hardly cause organizational change. For this to take place, a strong coalition of local actors has to support the cause. Interventions from TMNs have, as our cases demonstrate, been able only to partially contribute to small organizational changes. For example, even though Johannesburg has been an early member of the C40 network and received substantial support from the network for more than a decade, we can find only small changes in organization and policy. The case of São Paulo provides a second example to our argument. The city is a founding member of C40 and a longstanding member of ICLEI. However, a political change with a new mayor in 2013 brought to a halt all climate-related activities, and the city stopped its involvement with TMNs. The existing climate change organizational unit and legal framework were put on ice and more or less remained under inertia for the rest of the term. These examples clearly show the limits of TMNs effects on urban climate politics and should be carefully considered when portraying cities as the new arena for action on climate change.

Third, we find it almost impossible for TMNs alone to break through barriers in the domestic policy context. Very often, a traditional focus of governments on pro-poor development and the emphasis of the 'right to development' sets the pulse for urban climate politics, prioritizing developmental aspects over environmental concerns. Features of the political and administrative system further shape the potential of cities to implement change. Cities are bound to act within the range that the degree of political, administrative, and fiscal decentralization allows to them. Whereas Indian cities are strongly steered by national or state governments and have little capability to formulate own policies, cities in Brazil's federation are much more autonomous, but also prone to abrupt shifts in political preferences. While having many administrative competences, even cities in Indonesia, Brazil, and South Africa lack financial resources for advancing climate mitigation because of the set-up of intergovernmental fiscal relations.

Also, the gridlock situation in South Africa's energy sector, Indonesia's reliance on coal-based energy, or Brazil's subsidizing of diesel will not be resolved through the participation of local governments in TMNs alone, even if joint learning and a transfer of good practices occur. Depending on the sector, drivers or veto players and their intertwinement with vested interests of ruling elites shape how cities can act on climate change. However, co-benefits and cost-effectiveness can overhaul the dominance of old interests, as we see in the city-wide installation of LED street lamps or solar initiatives, for example in India. We also found that political leadership and agency can make a difference in the local context. International recognition and ownership of the mayor have influenced the course of events in Bogor, Cape Town, and Jakarta. Nonetheless, the positive impact of political leadership has to be treated cautiously: if a climate change agenda is too strongly attached to a personality, and weakly institutionalized, it may become vulnerable, and can easily be reversed by an incoming government. Thus, progress often depends on personal authority that in turn leads to highly clientelistic systems and might not be sustained, once the mayor has left office.

Our discussion suggests that the effects of TMNs on a country's climate policy are more limited than widely assumed. Even in an era of global climate governance, domestic economics, politics, and institutions matter a great deal and critically influence the carrying out of climate policy-making on the ground. There are no universal implementation templates and the influence of TMNs depends on the characteristics of economic, political, and administrative systems. In theoretical terms, this suggests that authors dealing with transnational actors in global environmental governance need to more robustly theorize their interlinkages to the domestic contexts and the multilevel dynamics within nation-states. We argue that overlooking their constraining or enabling nature can lead to missing opportunities to create stable and long-term structures to address climate change.

While a few studies have recently addressed this desideratum, a lack of systematic and in-depth comparisons limits the extent to which they are generalizable. Future research should take a stronger account of such linkages and include them into analytical frameworks instead of placing the focus solely on relationships between transnational networks and cities. Second, by connecting the analysis of transnational climate governance with the literature on policy change, important insights could be gained on the domestic conditions that enable or hinder external agency, and which forms of agency can lead to sustainable policy and organizational change.

In practical terms, we argue that policymakers concerned with the role of TMNs in the policy domain of climate change need to broaden their range of cooperation and concentrate some of their efforts on supporting local public administrations and enhancing cooperation among local stakeholders, in order to adjust local institutions towards sustainable climate policies and action. Without gradual institutionalization, climate action can be like a candle in the wind – if a mayor's office changes, political will ceases, and urban climate projects could be extinguished. Neglecting adequate institutionalization of climate policy at the local level will seriously constrain the expected contribution of cities for global emissions reductions, and lead to a rude awakening in times when action will be needed even more urgently than today. Finally, TMNs also increasingly need to interact with economic actors, and national and state governments to enable cities to take climate actions. Overall, cities will not rescue the planet as long as there is a lack of cooperation and coordination among governmental levels, economic actors, and civil society.

References

Acuto, Michele. (2013). The new climate leaders? *Review of International Studies*, 39 (4): 835–857.

Andonova, Liliana B., Betsill, Michele, & Bulkeley, Harriet. (2009). Transnational climate governance. *Global Environmental Politics*, 9(2): 52–73.

Anguelovski, Isabelle, & Carmin, JoAnn. (2011). Something borrowed, everything new: Innovation and institutionalization in urban climate governance. *Current Opinion in Environmental Sustainability*, 3(3): 169–175.

Betsill, Michele, & Bulkeley, Harriet. (2004). Transnational networks and global environmental governance: The cities for climate protection program. *International Studies Quarterly*, 48(2): 471–493.

Betsill, Michele, & Bulkeley, Harriet. (2007). Looking back and thinking ahead: A decade of cities and climate change research. *Local Environment*, 12(5): 447–456.

Bulkeley, Harriet. (2010). Cities and the governing of climate change. *Annual Review of Environment and Resources*, 35: 229–253.

Bulkeley, Harriet, & Betsill, Michelle. (2003). *Cities and Climate Change: Urban Sustainability and Global Environmental Governance*. London: Routledge.

Bulkeley, Harriet, Davies, Anna, Evans, B. O. B., Gibbs, David, Kern, Kristine, & Theobald, Kate. (2003). Environmental governance and transnational municipal networks in Europe. *Journal of Environmental Policy & Planning*, 5(3): 235–254. DOI:10.1080/1523908032000154179.

Busch, Per-Olof, Jörgens, Helge, & Tews, Kerstin. (2005). The global diffusion of regulatory instruments: The making of a new international environmental regime. *The Annals of the American Academy of Political and Social Science*, 598: 146–167.

Capano, Giliberto. (2009). Understanding policy change as an epistemological and theoretical problem. *Journal of Comparative Policy Analysis: Research and Practice*, 11(1): 7–31. DOI:10.1080/13876980802648284.

Carpenter, Daniel. (2010). *Reputation and Power: Organizational Image and Pharmaceutical Regulation at the FDA*, 1st edn. Princeton Studies in American Politics: Historical, International, and Comparative Perspectives. Princeton, NJ: Princeton University Press.

Chan, Sander, Falkner, Robert, Goldberg, Matthew, & van Asselt, Harro. (2018). Effective and geographically balanced? An output-based assessment of non-state climate actions. *Climate Policy*, 18(1): 24–35. DOI:10.1080/14693062.2016.1248343.

Dolowitz, D., & Marsh, D. (2000). Learning from abroad: The role of policy transfer in contemporary policy-making. *Governance*, 13(1): 5–23. DOI:10.1111/0952-1895.00121.

Dolowitz, David, & Marsh, David. (1996). Who learns what from whom: A review of the policy transfer literature. *Political Studies*, 44 (2): 343–357.

Fisher, Dana R. (2013). Understanding the relationship between subnational and national climate change politics in the United States: Toward a theory of boomerang federalism. *Environment and Planning C: Government and Policy*, 31(5): 769–784.

Fünfgeld, Hartmut. (2015). Facilitating local climate change adaptation through transnational municipal networks. *Current Opinion in Environmental Sustainability*, 12: 67–73. DOI:doi.org/10.1016/j.cosust.2014.10.011.

Garschagen, M., & Romero-Lankao. P. (2015). Exploring the relationships between urbanization trends and climate change vulnerability. *Climatic Change*, 133: 37–52. DOI:10.1007/s10584-013-0812-6.

Gordon, D. J., & Johnson, C. A. (2017). The orchestration of global urban climate governance: Conducting power in the post-Paris climate regime. *Environmental Politics*, 26 (4): 694–714.

Hall, Peter A. (1997). The role of institutions, interests and ideas in the comparative political economy of the industrialized nations. In Mark Zuckerman & Alan Lichbach (eds.), *Comparative Politics: Rationality, Culture and Structure*. New York: Cambridge University Press, 174–207.

Hickmann, Thomas. (2016). *Rethinking Authority in Global Climate Governance: How Transnational Climate Initiatives Relate to the International Climate Regime*. London: Routledge.

Hickmann, Thomas, Fuhr, Harald, Höhne, Chris, Lederer, Markus, & Stehle, Fee. (2017). Carbon governance arrangements and the nation-state: The reconfiguration of public authority in developing countries. *Public Administration and Development*, 37(5): 331–343. DOI:10.1002/pad.1814.

Howlett, Michael, Ramesh, M., & Perl, Anthony. (2009). *Studying Public Policy: Policy Cycles and Policy Subsystems*, 3rd edn. Don Mills, Ontario: Oxford University Press.

Kern, Kristine, & Alber, Gotelind. (2009). Governing climate change in cities: Modes of urban climate governance in multi-level systems. The international conference on Competitive Cities and Climate Change, Milan, Italy, 9–10 October, 2009.

Kousky, Carolyn, & Schneider, Stephen H. (2003). Global climate policy: Will cities lead the way? *Climate Policy*, 3(4): 359–372.

Lederer, M. (2018). External state actors. In T. Risse, A. Draude, & T. A. Börzel (eds.), *The Oxford Handbook of Governance in Areas of Limited Statehood*. Oxford: Oxford University Press, 191–210.

Lee, Taedong, & Koski, Chris. (2014). Mitigating global warming in global cities: Comparing participation and climate change policies of C40 cities. *Journal of Comparative Policy Analysis: Research and Practice*, 16 (5): 475–492. DOI:10.1080/13876988.2014.910938.

Lee, Taedong, & Van de Meene, Susan. (2012). Who teaches and who learns? Policy learning through the C40 cities climate network. *Policy Sciences*, 45 (3): 199–220.

Lodge, M. (2012). Administrative patterns and national politics. In B. Peters & Jon Pierre (eds.), *The SAGE Handbook of Public Administration*. London: SAGE, 545–560.

Peters, B. G., & Pierre. J. (2006). Introduction. In B. G. Peters & J. Pierre (eds.), *Handbook of Public Policy*. London: SAGE, 1–11.

Rashidi, Kaveh, & Patt, Anthony. (2017). Subsistence over symbolism: The role of transnational municipal networks on cities' climate policy innovation and adoption. *Mitigation and Adaptation Strategies for Global Change*. DOI:10.1007/s11027-017-9747-y.

Roberts, Debra. (2008). Thinking globally, acting locally – institutionalizing climate change at the local government level in Durban, South Africa. *Environment and Urbanization*, 20(2): 521–537. DOI:10.1177/0956247808096126.

Roger, Charles, Hale, Thomas, & Andonova, Liliana. (2017). The comparative politics of transnational climate governance. *International Interactions*, 43(1): 1–25. DOI:10.1080/03050629.2017.1252248.

Román, Mikael. (2010). Governing from the middle: The C40 Cities Leadership Group. *Corporate Governance: The International Journal of Business in Society*, 10(1): 73–84.

Rose, Richard. (1993). *Lesson Drawing in Public Policy: A Guide to Learning across Time and Space*. Chatham: Chatham House Publishers.

Selin, Henrik, & VanDeveer, Stacy D. (2012). Federalism, multilevel governance, and climate change politics across the Atlantic. In Paul F. Steinberg & Stacy D. VanDeveer (eds.), *Comparative Environmental Politics. Theory, Practice and Prospects*. Cambridge, MA: MIT Press, 341–368.

Streeck, Wolfgang, & Thelen, Kathleen. (2005). *Beyond Continuity: Institutional Change in Advanced Political Economies*. Oxford: Oxford University Press.

Thoenig, Jean-Claude. (2012). *The SAGE Handbook of Public Administration*. London: SAGE.

Toly, Noah J. (2008). Transnational municipal networks in climate politics: From global governance to global politics. *Globalizations*, 5(3): 341–356.

Valente de Macedo, Laura, Setzer, Joana, & Rei Fernando. (2016). Transnational action fostering climate protection in the city of São Paulo and beyond. *The Planning Review*, 52(2): 35–44.

13

The Politics of Urban Climate Futures

Recognition, Experimentation, Orchestration

JEROEN VAN DER HEIJDEN, CHIARA CERTOMÀ, AND HARRIET BULKELEY

Taken together, this volume constitutes a systematic mapping, exploration, and interrogation of the nature of agency and empowerment in urban climate politics and action. It brings together contributions from scholars with different disciplinary backgrounds whose research focus on a variety of geographical areas and political positions. The breadth and depth of these contributions – theoretically, conceptually, empirically, and methodologically – speak to the driving questions outlined in the introduction to the book: *who* are the novel agents in urban climate governance; *how* are they empowered; and *what* is the empowering effect of increased agents and agency in urban climate governance? In this brief concluding chapter, we reflect on these contributions and synthesize the main arguments presented. This leads to key lessons on agency and empowerment in urban climate politics and opens up challenging perspectives for a future research agenda on these themes in the politics of urban climate futures.

Which Agents Are Central to Urban Climate Politics?

Having been asked to reflect on the nature, features, organizational modes, and potentialities of agents emerging through urban climate transitions, we find that the contributing authors have focused on a bounded set of actors. Whether this is because of semantic differences or the more common-sense use of the term 'actors' as a means of reflecting on agency in the policy arena, the chapters are still clear that not all individuals or organizations have the capacity to shape the politics and direction of urban climate action. Here, we focus on the agents identified in the book as having this capacity – the agency to enable change. The chapters in this book point to three critical sets of actors that have such capacity.

The first are *translocal and transnational municipal networks* (TMNs – see particularly Chapters 2, 4, 11, and 12). What is of interest here are the essentially different capacities that the contributing authors to this book find in TMNs. In his contribution, Gordon (Chapter 2) is interested in the capacity that TMNs themselves have and the capacities they create. That is, so argues Gordon, TMNs are in a unique position to organize cites around a common goal and create or strengthen the capacities of the cities that are organized in TMNs to achieve that goal. The capacities created, or at the very least strengthened by TMNs, are central to the other three chapters that focus on TMNs. As an example, TMNs can help cities to recognize the capacities they have to act on climate change and to make climate action a central aspect of their urban politics, as was illustrated through a case study in Brazil by Barbi and Valente de Macedo (Chapter 4). Or, TMNs can develop tailored programmes and sub-networks for cities facing a similar set of problems within their membership base so that they can share resources and work together on addressing those problems, as was illustrated through a case study in Sub-Saharan Africa by Gore (Chapter 11). Yet, as explained by Stehle et al. (Chapter 12), while TMNs may have the capacity to generate or expand climate action by city governments, such climate action is by no means guaranteed to achieve its desired outcomes if it is not supported by actors other than city governments and TMNs.

A second dominant cluster of agents are *municipal governments and local public-sector agencies* (see particularly Chapters 3, 4, 6, 9, and 10). The rise – or perceived rise – of these actors in the global climate governance area is noteworthy considering long-standing critiques to higher levels of government for not being able to effectively act on climate change. While it remains to be seen whether city governments and local public sector agencies will be more successful than their national and regional counterparts, the collected chapters indicate that democratically elected local officials and publicly accountable organizations are a necessary (albeit not sufficient) aspect of meaningful urban climate action. For example, Patterson and van der Grijp (Chapter 3) illustrate how the Amsterdam City Government has used various capacities – financial resources, the power of senior city leadership, time of the civil sector – to develop, implement, and cultivate narratives of climate action to support its urban climate policy agenda. Yet, as became clear in Ninomiya and Burch's contribution on the Waterloo Region in Canada (Chapter 9), city governments may encounter limitations to their capacity to act to climate change that are comparable to those experienced by higher levels of government. That is, local interest groups may seek to block climate action, consensus-oriented decision-making processes may lead to watered-down policies rather than a radical trajectory of deep decarbonization, and the full range of economic and social impacts of any local urban climate intervention may be difficult to justify for elected city leaders.

A third and final dominant cluster of agents involved in urban climate politics are *citizens and communities* (see particularly Chapters 6, 7, 8, and 10). The scale at which citizens have begun to organize themselves in communities, grassroots organizations, and TMNs for urban climate action is impressive and well documented in the urban climate governance literature. A trend less well documented in this literature – and more central to urban studies scholarship – is the ongoing responsibilization of citizens in urban climate action as described by Hughes et al. (Chapter 7) and Uittenbroek et al. (Chapter 10). Such responsibilization appears a forced shift from governments-as-agents to citizens-as-agents, and goes against high expectations expressed about the capacity of city governments and local public sectors – and perhaps even TMNs as their binding and representative units – to effectively act on climate change. The handing down of responsibility for deciding and managing urban climate politics also attracts the critiques of radical geographers and political theorists who claim this to be a further example of ongoing privatization of public space and sphere in the context of neo-liberal governance (Harvey, 2005). While private actors (citizens as well as non-profit and for-profit organizations) take advantage of the void left by the retreat of public authority and control, others may lack the means to do so (Egyedi & Mehos, 2012). The contribution of Hughes et al. (Chapter 7) provides the clearest illustration here: where data – novel and archived – are more easily accessible than ever, they are still not equally accessible to everyone. This gives those with the capacity to create, store, and access data considerably more power in urban climate politics than those who lack this capacity.

In summary, the 'stock take' undertaken in this volume from the Earth System Governance (ESG) work over the past decade shows that the agents within urban climate politics are largely the same as in the decade before – mainly TMNs and municipal governments. Though, it appears that citizens and communities are becoming a more important agent at the urban scale also. The growing role of citizens and communities helps creating space for community action, but because of 'responsibilization' the increasing role of citizens and communities is seen in a critical light by the various authors also. Considering TMNs and municipal governments, we see a deepening of their roles, a strengthening of the capacities they have, and, most importantly, the capacities they are providing to other actors. That is, both TMNs and municipal governments have secured a central role in urban climate politics and we expect this role will only grow in the politics of urban climate futures. At the same time, we observe, through the chapters contributed by the authors in this book, that TMNs and municipal governments are increasingly providing others with capacities to act to mitigate and adapt to climate change. We turn to this in the sections that follow.

From Agents to Agency?

While the chapters in this book focus on the growing importance of urban actors in climate governance, they also provide insight into the shifting nature of agency being produced within cities to address climate change. We find distinctions drawn between individual and collective agency (Chapters 3), place-based and global agency (Chapter 2), and fragmented or weak versus united or strong agency (Chapter 11). To the critical reader of the chapters, this may illustrate the difficulty of analytically conceptualizing and empirically capturing the novel forms of agency, whose overlapping capacities do not necessarily lie with a specific agent, and whose boundaries are often blurry. In our reading, the chapters demonstrate that individual agents often do not have sufficient capacity to act – for example, because they lack resources, or their desired actions are resisted by others – and that meaningful forms of agency are created in the interplay of the capacities of different actors. Three such forms of collective agency stand out in the various chapters.

The first of these is the collective agency that TMNs and cities have to *pursue and provide recognition* of cities as important agents in addressing climate change (see particularly Chapters 2, 4, 11, and 12). In practical terms, and as illustrated in great detail in Gordon's contribution (Chapter 2), TMNs make claims to cities' global leadership in responding to climate change; generate performance measurement standards for cities to report the impact of their actions and to compare themselves with others; set codes of conduct for their member cities to adhere to; and motivate cities to develop and implement climate action, among others. As Gordon illustrates further, these actions have helped cities to gain recognition in the global climate regime and allowed TMNs to position themselves – pursue recognition – as essential actors in the global climate regime also. Yet, as the contributions by Barbi and Valente de Macedo (Chapter 3) and Gore (Chapter 11) indicate, TMNs have only a limited impact on the climate actions taken by their member cities. Albeit observed in different contexts (Brazil and Sub-Saharan Africa respectively), if these member cities lack the capacity to act to mitigate and adapt to climate change or if their actions are contested or resisted by other actors, it is unlikely that such actions achieve desired results. Eventually such a lack of effect of TMN-inspired climate action on the ground may backfire and may raise questions about the necessity of TMNs in the global climate regime. Here it becomes clear that TMNs need (active) cities to gain and maintain recognition just as much as cities need TMNs. This may provide one explanation for the long-standing observation that such networks tend to focus on 'leaders' rather than ordinary cities or those who lag behind in taking action on climate change (see Kern & Bulkeley, 2009).

A second form of collective agency pointed out by the contributors is *innovation and experimentation* (see particularly Chapters 3, 4, 8, and 9). Echoing expectations on innovation and experimentation in the broader climate governance literature (Hoffmann, 2011), the contributors to this book are particularly hopeful about this second form of agency because it provides lessons on what kind of urban climate action yields desirable outcomes, where and why, and whether it can be scaled up or out. They are, however, also critical and illustrate that innovation or experimentation by itself is no guarantee for meaningful and effective urban climate action. We observe, again, that the strongest agency is expected at the interplay of the capacities of various actors involved in processes of innovation and experimentation. For example, Mancebo and Certomà (Chapter 8) illustrate that innovative forms of governance for urban agriculture are most likely to yield desired outcomes when they bring together the capacities of communities, professionals, and other stakeholders with those of urban planners and public officials. Communities may hold tacit knowledge that planners lack; professionals may be able to extract this tacit knowledge from individuals because of trust relations they have built over time; planners may be able to translate such tacit knowledge to workable plans in ways that communities and professionals cannot; and so on. The Decarbonize Waterloo Region case presented by Ninomiya and Burch (Chapter 9) echoes a similar finding: 'one-way' approaches to innovation or experimentation led by already powerful stakeholders will likely face resistance, exclude minority groups, and underuse the potential of stakeholders to achieve deep decarbonization.

The third form of collective agency discussed in the chapters is the *orchestration* of urban climate action (see particularly Chapters 3, 6, 10, and 11). Such orchestration involves the mobilization of various actors and by an agent in the pursuit of a joint goal. The agent normally cannot command the actors to act towards this goal, but can incentivize them doing so through material or ideational support (see further Abbot, Genschel, & Snidal, 2016). As with experimentation, orchestration is not located with a specific agent – TMNs, city governments, and citizen alliances are all found to orchestrate urban climate actions. The contributions of Patterson and van der Grijp (Chapter 3) and Uittenbroek et al. (Chapter 11), for example, are illustrative of how city governments in the Netherlands are mobilizing various actors to take a range of climate actions that all contribute to larger local urban climate goals. The energy transition of Amsterdam is illustrative here. Rather than fully mapping out this transition by itself, the Amsterdam City Government has in place a framework policy – the City Deal – that welcomes initiatives from citizens and local firms. Through financial and other support the Amsterdam City Government can back those initiatives that fill essential gaps or provide critical knowledge or expertise to achieve its energy transition. In doing so, the Amsterdam

City Government makes it clear that it lacks the capacity to achieve this transition by itself and needs others to achieve it. As Uittenbroek et al. rightfully notice, however, when governments take such a step back and call on its citizens and local firms to support them in achieving certain climate goals there are risks of maladaptation when the desired support is not delivered, and citizen fatigue when they feel overcharged with responsibilities. Drawing on examples from the metropolitan areas of Accra and Berlin, Schulz and Bruns (Chapter 6) also illustrate that the orchestration of urban climate actions is increasingly shaped by the agency of new cross-movement alliances that are simultaneously operating as transnational policy networks, local advocacy groups, and citizen-led protest movements, while bringing together actors from diverse political, environmental, rights-based, and spiritual backgrounds.

In summary, looking over the contributions, it becomes clear that the actors we are used to in urban climate politics have had to create novel forms of agency to achieve their desired goals – and are still in the process of creating such novel forms of agency. These novel forms of agency work in-between formal kinds of capacity such as legitimacy and power, and traditional resources such as finance and staff, and appear most promising when they work at the interplay of the capacities of different agents. These three forms of collective agency – recognition, experimentation, and orchestration – capture the most important and novel contributions made over the last decade by ESG scholars and will, we feel, influence the thinking of and research on urban climate politics in the next decade to come. Of interest for future scholarship will be the interdependencies of actors, and their possibilities to mutually reinforce each other's capacities. Carefully unpacking actor constellations to understand how individual capacities combine and what configurations of actors and capacities result in meaningful urban climate action will be essential to provide timely advice to policymakers and practitioners involved in urban climate transitions.

From Agency to Empowerment?

Two broad and somewhat contrasting conclusions emerge from the chapters on the question whether this increased agency has resulted in increased empowerment. A set of chapters concludes that increased agency does not necessarily result in increased empowerment also. This insight is robustly confirmed by, among others, Barbi and Valente de Mancebo (Chapter 4) and Gore (Chapter 11), who indicate that while TMNs have increased the agency of cities in Latin America and Africa to develop climate actions, a change in power relationships between cities and other authorities still is required to achieve the goals of these actions. In short, increased agency here does not imply increased empowerment. Hughes et al. (Chapter 7), in their contribution, illustrate how the availability of (big) data may increase the agency of agents and empower some, but not others. Data can empower new agents to engage in urban climate action, they can further empower those already engaged by broadening their power-base, and they can empower agents through opportunities of participation in the creation of data. Yet, the very agents who create and control data may disempower others for instance by means of data availability, or use data standards or formats in a way that disempowers marginalized groups who do not conform to those formats or standards. This dual side of empowerment through a specific form of agency was also illustrated by Uittenbroek et al. (Chapter 11), who indicate that the responsibilization of citizens may empower some, but not others because of frictions, maladaptation, and an inability of some citizens to act on their assumed responsibilities. This set of chapters is critical to the (limited) extent that this increased agency has empowered the actors involved in urban climate politics and governance.

Yet, another set of chapters has a more hopeful outlook. Their authors point out that this increased agency has resulted in increased empowerment of urban climate politics and governance overall. They are aware that the changes in empowerment perhaps not achieve the necessary urban climate transitions within the timeframe available, nor that all actors in urban climate politics have been empowered equally or desirably. The new situation of increased empowerment is, however, they point out, preferable to the former situation of lesser empowerment overall. This becomes clearest in the chapters that explore how some actors create capacities or strengthen capacities of others. For example, Gordon's contribution (Chapter 2) illustrates how TMNs have pursued recognition of cities overall as central actors in the global climate regime. Through this recognition, cities find it easier to justify the climate actions they are taking, find it easier to work together, and so on. In similar vein, the interest of municipalities in seeking innovative solutions to urban climate problems through experiments has, overall, empowered citizens, private actors, universities, and others to bring forward their ideas and desires for climate action and sometimes see these taken up. For example, the mere framing of some of these experiments as 'urban laboratories' (discussed by Patterson and van der Grijp, Chapter 3) justifies the actions taken by the actors working in these – whether they are successful or not – and empower them as a collective. In sum, this set of chapters looks at the novel forms of collective agency discussed before as empowering urban climate politics and governance in achieving desirable urban climate futures (empowerment as a process), rather than as empowering the actors of and affected by urban climate politics (empowerment as an outcome).

The various chapters also point out that making visible the interplay between increased agents and agency, and increased or decreased empowerment is all but easy. Processes of empowerment and disempowerment are a complex interplay of agents, the institutional resources available, and barriers they face. The same applies to the pursued urban climate action goals. Ninomiya and Burch (Chapter 9) aptly consider this as a black box that is difficult to open or understand. To better understand processes of empowerment and disempowerment and dynamics of inclusion and exclusion, they propose that scholars become more involved in 'learning by doing', for instance through participatory action research. Deep case studies are of utmost importance, we acknowledge, because these allow for a deep understanding of specific processes within a determined context, in consideration of the evidence that the knowledge base of the empowering effect of increased agents and agency in the politics of urban climate futures is very limited. Yet, we are wary of a future knowledge base that is fragmented and dispersed and does not allow for drawing lessons across cases. Here we are hopeful about the heuristic framework proposed by Patterson and van der Grijp (Chapter 3) to make processes of empowerment and disempowerment not only visible, but also comparable across observations. Their framework considers empowerment – defined as 'the process of enhancing the capability of a collective initiative to realize a desired outcome' – as the result of interplay among agency, opportunity structures, and shifts in power. The combination of these three core conditions allows for a more dynamic conceptualization of empowerment or disempowerment, compared to the one offered by the capacity approaches, as it moves beyond the availability of resources available to agents and barriers they may face. It effectively questions how agents make use of these resources, how they seek to address barriers, and how endogenous and exogenous factors have actively or passively helped changing power structures or prevent this to happen.

Moving beyond the detailed analytical and methodological challenges addressed in the chapters, we see two broader ones – both are about making visible empowerment in studies of urban climate politics. First is the challenge of focus and use of the concept 'empowerment'. From the chapters it has become clear that some conceptualizations of empowerment in the politics of urban climate futures are easier to trace in empirical studies than others. For example, where empowerment is seen in terms of the inclusion of specific actors in urban climate transitions it is easier to trace than where it is seen in terms of creating the capacity to enable social justice outcomes of urban climate transitions. From the chapters, it has become clear also that the concept 'empowerment' has not been developed in urban climate politics to a similar degree as it has been, for example, in urban studies (Gottdiener, Budd, & Lehtovuori, 2016; Lyons, Smuts, & Stephens, 2001), legal scholarship (Banik, 2011; Golub, 2010), or sociology more generally (Oakley, 2001; Pease, 2002; Rowlands, 1997; Wilkinson, 1998). Rather than developing their own conceptualizations, scholars of urban climate politics may wish to build on those developed in other disciplines for their empirical studies and find inspiration there on how to put this inherently contested concept to practical use. A second challenge is methodological: how can we make visible the complex interplay among agents, agency, and empowerment? Building on the chapters, we particularly see merit in configurational methods (Fiss, Cambré, & Marx, 2013; Rihoux & Ragin, 2009). The chapters all point out that empowerment is a result of various conditions or factors working together. Configurational methods take as a point of departure that it is often not an individual factor (say, the agency of a TMN) that explains an outcome best (say, the empowerment of a city), but that various factors combine in achieving that outcome with none of them being sufficient to do so by itself. In addition, some configurational methods take as a point of departure that there is no single path or trajectory to a specific outcome, but that various paths or trajectories can all lead to the same outcome. Methods such as Qualitative Comparative Analysis (Rihoux & Marx, 2013, for an application see Van der Heijden, 2017) and causal process tracing (Collier, 2011; for an application see Biesbroek et al., 2014) are, we feel, ideally suited to unpack the black box of empowerment in urban climate governance in a systematic, transparent and replicable way, while allowing for cross-case comparisons that may help to trace patterns that could hold more generally, beyond the cases studied.

Future Research and Implications for Policy and Practice

The contributions to this book have provided a window through which the scholarly engagement with agency and empowerment in the politics of urban climate futures can be observed. It will come as little surprise that we here call for increased future research in this field, and deeper engagement with these issues (Rickards reaches a related conclusion in Chapter 5). More specifically, research on increasing agency and contested empowerment in the politics of urban climate futures needs to recognize the various understandings of the very terminology it uses.

In our introductory chapter (Chapter 1), we suggested a broad definition of agency characterized by the propensity of social, socio-material, and socio-natural collectives to produce collective processes leading towards a goal that – despite not necessarily linguistically negotiated – nevertheless represents the emerging outcomes of single agents' desires. Despite the increasing interest in scholarly research for the political effects of heterogeneous formations' agency, our broad definition resonated less well with the novel agents studied and the case studies presented in the various chapters – as a result it was not further developed in this book. Rather, across the chapters, contributors have stayed close the narrow definition of agency that we presented in the introduction to this book – that is, the capacity of individuals or organizations ('agents') to act independently and autonomously towards achieving desired outcomes. This leads us to two reflections. First, even where such a relatively conventional approach to agency is adopted, we see that there are critical agents of change absent from the analysis – particularly those who have traditionally been regarded as operating in the 'private' sphere, such as businesses or investors, but also a wide array of agents who are essential to the life and work of cities, from churches to schools, non-governmental organizations to clubs. Whether this absence reflects a methodological bias towards municipal agencies as standing in for 'the city' or reflects the absence of such actors from an engagement with urban climate politics and action, further research is needed to tease out the agents of change that need to be engaged within cities in order to move this agenda forward. Second, the focus on conventional forms of agency seems to indicate that the broad ESG community is traditionally more interested in human and organizational agency than in non-human or more-than-human agency. This may explain why the dominant novel agents identified in this book are TMNs, city governments and local public sectors, and citizens – and why, strikingly, little attention is paid in the various chapters to non-human and more-than-human agents. While this may reflect the wider disciplinary backgrounds of the ESG community, it does not fully encompass the range of research that is now being conducted on what it means to live an urban life in the Anthropocene, how we should think about the nature and dynamics of urban infrastructure, or where and how forms of transformation to more sustainable futures can be found. Expanding the conceptual repertoire of the ESG community engaging with urban climate change may require both reaching out to new disciplines and enabling those within the community to develop novel perspectives on the city and its governing.

In similar vein, the dominant forms of increased agency observed relate to these human and organizational agents: pursuing and providing recognition, innovation and experimentation, and orchestration. These are all fairly high-level and visible forms of agency. While less visible and less formal forms of agency – likely based in bottom-up processes of empowerment – and their empowering and disempowering effects are more difficult to explore, understanding them is of essence. Combined, these may be the friction required to cause systemic changes in the politics of urban climate futures and set it on a pathway towards achieving urban futures with greater well-being of individuals and societies. To better understand increasing agency in this area and its empowering effects, more historical accounts of novel agents, novel forms of agency, and empowering and disempowering processes are required. It is essential also that scholars expand the contexts they study as well as the agents they are interested in – thus, a move is required beyond the conventional TMNs, cities, and local action communities we see recurring time and again in current scholarship. This also requires more engagement theoretically and conceptually with whether the forms of increased agency have also increased the capacity to transform urban futures in a low-carbon direction and address climate change. It is vital that we do not equate the development of capacity with the creation of empowerment to achieve change on the ground. Equally, even where empowerment to act is forthcoming, its consequences in terms of realizing socially and environmentally just urban futures are not secure. Further work is required to understand how generating capacity and empowerment to govern climate change in the city may create unintended consequences for vulnerable communities and environments, or fail to ensure that those that bear the most responsibility for action are called upon to act. More practically, we see a need for more direct engagement between scholars and policymakers, action communities, and other local climate leaders to help increasing agency and help empowering urban climate action.

These insights have implications for policy and practice also. For policymakers the main insights from this book are that merely adding more agents to local climate governance regimes, or merely giving more agency to existing ones is no guarantee for desirable climate action to emerge. In similar vein for practitioners, merely demanding or taking more agency in urban climate politics is no guarantee that urban climate action goals will be achieved. Without a careful consideration of the interactions between the institutional resources agents have available, the institutional barriers they face, and the goals they pursue, increased agency may not yield increased empowerment – or, worse, may result in disempowerment. In such situations, added agency might be even more toxic than a lack of agency as it may create a downward spiral of bitterness and disillusionment. This book has sought to make visible how such downward spirals can be prevented, and how adding agents and agency may improve empowerment to ultimately increase the quality and quantity of urban climate action.

References

Abbot, Kenneth W., Genschel, Philipp, & Snidal, Duncan. (2016). Two logics of indirect governance. *British Journal of Political Science*, 46 (4): 719–729.

Banik, Dan, ed. (2011). *The Legal Empowerment Agenda: Poverty, Labour and the Informal Economy in Africa*. Farnham: Ashgate Publishing.

Biesbroek, G. Robbert, Termeer, Catrien, Klosterman, Judith, & Kabat, Pavel. (2014). Rethinking barriers to adaptation. *Global Environmental Change*, 26(May): 108–118.

Collier, David. (2011). Understanding process tracing. *Political Science & Politics*, 44(4): 823–830.

Egyedi, Tineke, & Mehos, Donna, eds. (2012). *Inverse Infrastructures Disrupting Networks from Below*. Cheltenham: Edward Elgar Publishing.

Fiss, Peer, Cambré, Bart, & Marx, Axel, eds. (2013). *Configurational Theory and Methods in Organizational Research*. Bingley: Emerald Group Publishing.

Golub, Stephen. (2010). 'What is legal empowerment?' In Stephen Golub (ed.), *Legal Empowerment*. Rome: International Development Law Organisation, 9–18.

Gottdiener, Mark, Budd, Leslie, & Lehtovuori, Panu, eds. (2016). *Key Concepts in Urban Studies*, 2nd edn. London: SAGE.

Harvey, David. (2005). *A Brief History of Neoliberalism*. Oxford: Oxford University Press.

Hoffmann, Matthew. (2011). *Climate Governance at the Crossroads*. Oxford: Oxford University Press.

Kern, Kristine, & Bulkeley, Harriet. (2009). Cities, Europeanization and multi-level governance. *Journal of Common Market Studies*, 47(2): 309–332.

Lyons, Michal, Smuts, Carin, & Stephens, Anthea. (2001). Participation, empowerment and sustainability. *Urban Studies*, 36(1): 59–72.

Oakley, Peter. (2001). *Evaluating Empowerment*. Oxford: INTRAC.

Pease, Bob. (2002). Rethinking empowerment. *The British Journal of Social Work*, 32(2): 1.

Rihoux, Beniot, & Marx, Axel. (2013). QCA, 25 years after 'The Comparative Method'. *Political Research Quarterly*, 66(1): 167–235.

Rihoux, Beniot, & Ragin, Charles. (2009). *Configurational Comparative Analysis*. London: SAGE.

Rowlands, Jo. (1997). *Questioning Empowerment*. Oxford: Oxfam.

Van der Heijden, Jeroen. (2017). *Innovations in Urban Climate Governance*. Cambridge: Cambridge University Press.

Wilkinson, Adrian. (1998). Empowerment. *Personnel Review*, 27(1): 40–56.

Index